TO THE EDGES OF THE UNIVERSE

TO THE EDGES OF THE UNIVERSE

Space Exploration in the 20th Century

Don DeNevi

CELESTIAL ARTS
Millbrae, California

Photograph and illustration credits

NASA: Pages 5, 11, 16, 20, 28, 29, 30, 40, 48, 57, 58, 61, 62, 65, 67, 73, 84, 93, 97, 101, 104, 111, 114, 118, 121, 122, 133, 140, 141 (bottom), 143, 144, 147, 148, 155, 160, 164, 172.

Rockwell International's Space Division: Pages 21, 22, 23, 24 (top), 25, 27.

Lockheed Missiles & Space Company: Pages 24 (bottom), 41, 99, 100.

Boeing Aerospace Company: Pages 47, 49, 50, 52, 53.

Martin Marietta Aerospace: Page 141.

Cover design by Betsy Bruno
Cover art by Walter Zajowski

Published by CELESTIAL ARTS, 231 Adrian Road,
Millbrae, California 94030

First printing: February 1978
Manufactured in the United States of America

Library of Congress Cataloging in Publication Data

DeNevi, Donald P.
To the edges of the universe.

Bibliography: p.
Includes index.
1. Outer space—Exploration. 2. Space colonies.
3. Space sciences. I. Title.
QB500.D43 919.9 '04 77-90005
ISBN 0-89087-184-1
ISBN 0-89087-212-0 pbk.

1 2 3 4 5 6 7 8 83 82 81 80 79 78

CONTENTS

In friendship, this book is dedicated to
Maria Christina Igoa

Acknowledgments

The real authors of this book are the many space scientists, engineers and project staff personnel, inside and outside NASA, whose incredibly complex concepts were made available to me in simple, easy-to-understand explanations. For example, Grumman Aerospace scientists Ross Fleisig, Joel Bernstein, Gary Geschwind, Joe Marino, and Al Nathan, as well as Aerospace Corioration's Ivan Bekey and Harris Mayer, and Martin Marietta's George Alexander, John Gurr, and Robert Powers, along with NASA's Richard Fimmell, William Swindell, and Eric Burgess are especially to be credited for the quantity and quality of their writings. Prolific in technical detail, the writings and reports which make up much of this volume offer the public the latest and most advanced concepts man has yet come up with for the humanization of space.

Their sources have been supplemented by official NASA documents, priorities, and correspondence, along with information from private aerospace companies pertaining to their exploration activities in the outer fringes of the universe. Thus, in planning this volume, I have been obliged again and again to call upon the experts for advice. It is a pleasure now to acknowledge their assistance. All helped in the early discussions of the various projects as well as patiently answered specific questions.

I especially owe a great deal to the assistance and sympathetic cooperation of Dick Barton and Donald Patterson of the Rockwell International Space Division; George Butler of the McDonnell Douglas Astronautics Company; Bill Stewart of the Lockheed Missiles and Space Company;

Don Bane and Bill Becker of the Jet Propulsion Laboratory at the California Institute of Technology; Warren G. Lamb, Aerospace Corporation; Bill Mayes of the George C. Marshall Space Flight Center; Bill Rice of Boeing Aerospace Company; Pete Bertolino and Ruth Thomas of General Electric Company; Burnham Lewis of Grumman Aerospace; Frank Newsum and Melodie de Guibert of the Martin Marietta Aerospace Corporation; Mike Dunohoe and Mary Morash of the Ames Research Center; Vickie Pitts of the National Space Science Data Center; Tom Carpenter of the Goddard Space Center; Richard Piscara of the Stanford Linear Accelerator Center; John Newbauer, editor-in-chief, *Astronautics and Aeronautics;* Trudy Tiedemann and Hugh L. Dryden, Flight Research Center; William Lofard and Iris Scott, Lyndon Johnson Space Center; Jesse Crice, Strategic Aerospace Division, Vandenberg Air Force Base; William Rose, Superintendent of Planning Service, U.S. Government Printing Office; Patricia Andrews, Editorial Assistant, Grumman Aerospace *Horizons,* who so kindly sent artist transparencies on space solar power; Pat Oaks, National Geographic Society; and Bill O'Donnell, Captain Freitag, and especially Jesco Von Puttkamer of the National Aeronautics and Space Administration.

I am also indebted to personnel of the Lunar and Planetary Laboratory and Optical Sciences Center at the University of Arizona for astronomical information and photographs. And, finally, very special gratitude is extended to Stanford Linear Accelerator graphic artist Walter Zawojski who created the design for the book's jacket.

Preface

Is the earth really doomed in as few as fifty years? Advanced computers from the Massachusetts Institute of Technology answer in the affirmative after intensive analysis of this planet's expanding populations and diminished resources. Scientists, however, say "no" and point to the possibility of new self-contained celestial cities in outer space as one solution to the earth's problems.

The purpose of this book is to provide a simple, easy-to-read summary of possible opportunities for the future and the worthwhile human purposes these can serve, the goals and objectives developed to guide scientists toward these purposes, approaches being used to achieve these objectives, and the goals of this exploration.

In short, this volume is not NASA's plan for the future. Rather, it is an account of many of the ideas generated inside and outside of NASA. And, with appreciation for a difficult task well executed, I would like to recognize and recommend the following superb articles for the reader's further reading:

"Space Solar Power—An Available Energy Source," Grumman Aerospace *Horizons Magazine,* Vol. 12, No. 2, 1976.

"Coming: Permanent Occupancy of Space," *Horizons Magazine,* Vol. 13, No. 1, 1977.

"1980–2000: Raising Our Sights For Advanced Space Systems," *Astronautics and Aeronautics,* July/August, 1976.

The challenge NASA now faces is to select those key missions and critical programs that offer the greatest meaning to human life. Already scientists have begun transforming their studies into plans, and plans into realities. No plan, however, is as fantastic as the design for space settlement. Scientists are urging that the plan, costing as much as $200 million, be turned into reality within the next few decades. By that time, they warn, the option of moving into space may no longer be open to man.

1

The Promise of a New Era

I object to people running down the future. After all, I'm going to live the rest of my life there.

Thomas Edison

For most, the future always seems to be arriving a little before they are ready to give up the present. Not so for the National Aeronautics and Space Administration nor its hundreds of subsidiary aerospace companies. Neither downplaying the future, nor ignoring it, NASA is instead preparing for the health and economic security of some 200 million Americans. No one, of course, can really tell what will happen during the years 1980–2000. But whatever does happen, NASA is paving the way with the promise of a new space era.

If one were to read an astronomy textbook published before 1960, he would learn that virtually nothing was known about the surfaces and atmospheres of Venus, Jupiter, and Mercury; that the manned exploration of the moon was still 50 years away; and that scientists were still debating how the Martian "canals" came into being. Furthermore, earthbound astronomers and scientists had little to add to planetary planning projections.

But since the first small satellites were launched into orbit some eighteen years ago, the space program has made remarkable progress. Spaceprobes have now flown past Venus and Mercury, and many more are planned. An Orbiter has landed on Mars and discovered that no life forms as we know them exist on the mysterious red planet. Modern space-age Magellans have reached the moon, the very first earthlings to land and walk and jump in the world beyond earth.

Indeed, the NASA achievements have provided a new vision of the universe, one with glimpses into the nature of infinity and eternity that stagger the imagination. During the past few years, orbiting spacecraft as well as ground-based astronomy have brought to man's attention heavenly bodies, sources of energy, and stellar and galactic phenomena the nature and origin of which this world's scientists can only surmise.

Today, a great deal of work is being done by NASA to define future opportunities in space and design the hardware for those aeronautics. Of crucial importance is the answer to how man might best pursue those opportunities, the worthwhile purposes that America's future space endeavors and aeronautical research might serve, and the program objectives that can guide scientists toward fulfilling these purposes. NASA's recent achievements include Skylab with its incredible engineering and scientific breakthroughs, the Earth Resources Technology Satellite which has yielded tremendous information on the world's resources, Pioneer 10 which opened new horizons in man's understanding of the solar system and universe as it flew past Jupiter, and Viking 1's fantastic landing on Mars. Similarly, continued research in aeronautics is enabling the creation of military aircraft with greater combat effectiveness.

Says Homer E. Newell, NASA's associate administrator:

> As space scientists look to the future, they envision the possibility of centuries of continuing activity in space and aeronautics that can make many additional worthwhile and consequential contributions to our nation and the fulfillment of its objectives. In space, we envision a broadening of our opportunities during the 1980s made possible by the space shuttle. By providing economic, convenient, and reliable access to space, the shuttle stands as the keystone of our future capability to use space effectively for the benefit of man. Consequently, a great deal of thought has centered on how to utilize the new capabilities of the shuttle for beneficial purposes. In aeronautics, we see a broad spectrum of future opportunities made possible by an imaginative continuing program of aeronautics research.

Over the past twenty-five years, operational space programs have been replacing research and development programs, carrying with them rich dividends on all the money invested—and more riches are within grasp. Achieving them requires two crucial complex space systems: a long-endurance manned space colony and a low-cost space transportation system. The shuttle answers beautifully the low-cost need for traveling to space. The settlement facility would extend orbital stay-time and reduce the cost of operations even further.

The colony/station orbiting the planet earth has evolved from Salyut, Skylab, and Apollo-Soyuz. Combined with Spacelab, shuttle would allow for low-earth-orbit flights (sortie missions) of up to sixty days. Of the seven crew members, two would man the Spacelab. A while later, automated as well as manned free-flying Spacelabs would remain in orbit up to one hundred and twenty days. A space tug would tow them in conjunction with the shuttle. Serving as the first real earth-orbiting stations, the labs would be refitted and reflown after retrieval.

From that point on, the chronology of man's activities

in earth orbit should follow simply and progressively. The next project would see a small platform constructed from lightweight, flexible modules lifted into orbit and served by shuttle. The module would house twelve crewmen for up to eight months in zero-g (zero gravity). The crew would conduct a series of experiments as well as repair, construct, and launch additional satellites and spacecraft. Naturally, automated satellites and spacecraft would still be needed and would function for earthlings while the orbiting platforms were being created. The satellites would conduct experiments that best were handled by automated equipment and could be adversely affected by the environment of manned vehicles.

On July 17, 1975, when spacecraft from America and the Soviet Union joined together 138 miles above the earth, news stories around the world focused on the immensely complex technical feat of docking the modules and the implications of such successful coordination and cooperation by the two rival super powers. But almost completely ignored in press coverage of the Apollo-Soyuz mission was the incredible, fast chronology of man's activities in earth orbit that paved the way. Nearly every step of the way required not only creative thinking, but also new and brilliant engineering technology.

The Soviets launched the first crude earth-orbital station on April 14, 1971, into a 200-by-200 orbit with a 51.6° inclination. Scientists from the USSR termed it Salyut. But during the initial attempt to dock with it, a three-man crew from a Soyuz spacecraft could not unlatch its hatch. Nonetheless, the Soyuz 11 crew stayed in their spacecraft for some twenty-four days conducting stellar astronomy, agronomy, communications, and medical experiments. On their reentry, the entire crew was killed when a simple mechanical failure caused a rapid decompression. After that, the Russians made no further attempt to launch and man the Salyut. By controlled reentry, they destroyed it in October of 1974.

Apollo-Soyuz test project.

Skylab: "The Ugly Duckling of Space."

The Salyut vehicle was engineered to accommodate four men within a small space (84 cubic meters). The scheme included clustering four additional Soyuz around it in order to establish a prototype station. The station would have had four compartments: a forward docking and transfer tunnel derived from the Soyuz, a major control center, a small laboratory, and a maneuvering module.

America's Skylab followed Salyut into orbit in May of 1973. Launched by a single, two-stage Saturn V from the Kennedy Space Center, it was termed by many scientists "the ugly duckling of space" because it had one solar "wing" missing and had a makeshift sunshade replacing a lost heat shield. "But what a stately swan of science this 100-ton bird became," said Donald Hearth, NASA's Study Director. Maintaining the vehicle in repair with daring space walks, Skylab crews dramatically launched the post-Apollo era by providing man his most important studies yet of both sun and earth from space. Tens of thousands of incredible photos will yield information for a decade to come. Because of Skylab, scientists have already discovered new oil and ore deposits around earth, as well as new water sources for west Africa. Her 90 million-mile odyssey also provided a wealth of information about man's capabilities to operate in space. Among the achievements by Skylab and her crew:

- Most revolutions around the earth: 1,213
- Greatest distance flown: 90 million miles
- Longest successful repair demonstration outside craft: 7 hours and one minute
- Most data taken of earth: 20,500 photos

In total, Skylab's three missions devoted 515 man-days to orbital space experiments over a nine-month span.

However, as significant as these achievements were, the most important was the Apollo-Soyuz success during the late summer of 1975. Back in 1971, scientists from Russia and the United States began planning a docking mechanism to join spacecraft. A year later, a flight plan was drawn up for linking the vehicles for two days. The two nations constructed and tested the docking equipment as well as a crew-transfer module. The major objective of the mission was to determine the technical requirements for rendezvous, docking, and crew transfer for future manned stations and how well the equipment and procedures devised met them. This objective met with tremendous success.

There is little question today that the Salyut and Skylab successes of the '70s will give way to sortie missions by 1980. And, by the year 2000, hundreds of automated spacecraft for earth-resource monitoring, communications, navigations, and celestial observations will be orbiting the earth. But what will motivate future space activity?

Generally speaking, social, economic, scientific, political, and national security needs move people and governments to act. Since, hopefully, the United States and Soviet Union are entering an era of international cooperation in space, political and national security needs appear to be less important than before. Of real importance to mankind are those social benefits which offer an improved quality and style of life, economic benefits which bring monetary gains to people, and scientific benefits which increase man's understanding of the solar system, universe, and the laws governing them. In preparing an article on "Earth Orbiting Stations" for a NASA Space Systems Committee Report, Don Patterson et al, wrote:

> To date, the major benefits from the U.S. space effort have been scientific. Astronomy, solar physics, and exploration have made tremendous gains. Astronomers have located and categorized a large number of ultraviolet, X-ray, and gamma ray sources, discovered quasars and pulsars, and developed new and sometimes controversial theories on the origin, shape, and

> laws of the universe. Research in solar physics has spawned hypotheses on the structure and mechanisms of the Sun and its relation to other astronomical objects. Research in space physics has led to the discovery of the Van Allen belts, extended ionosphere, hydrogen geocorona, plasma pause, magnetopause, bow shock, geomagnetotail, trapped radiation, solar wind, and interplanetary magnetic field. Exploration of the Moon and planets has returned a wealth of new scientific information. The breadth of scientific advances from the first 18 years of space activity probably has no parallel in any comparable period in the history of mankind. Unfortunately, these advances take so long to affect the life of the man on the street that he has not been enthusiastic about them. The forcing function for continued space activity will not likely be scientific.

Thus, a whole new view of earth is necessary to cope with all the problems of population growth, dissipation of land, minerals, and water, as well as an environment that is slowly deteriorating. There is little question that earth observations must define the earth's geometry, surface characteristics, and dynamic body properties, and lead the way toward understanding the physics of the atmosphere, climate prediction, weather modification, and management of our resources and environment.

Such economic and social benefits as the following will add to those already flowing from automated earth-orbital spacecraft: Communications satellites make possible a global communications network at a large savings over ground-based equipment; meteorological satellites gather cloud-cover data which improve the accuracy and range of weather forecasts.

Capturing solar energy from orbit would alleviate the energy crunch and orbit observations would predict catastrophic earthquakes. Without question it is social motivations which are demanding environmental protection, but it is the economic benefits that will finally make it happen. To all concerned, the major justification for space activity during the next several decades is the social and economic benefits which will come from looking back at earth from space.

A SIMPLE PERSPECTIVE

In order to evaluate the need and potential for space activities and exploration for the remainder of the century, it might be wise to review past events of the space program. NASA officials point to four periods of space-related activity which seem to have had the greatest impact on American life: the years 1945 to 1957; 1957 to 1961; 1962 to 1969; and the decade of the '70s.

Even though the concept of flying through space was developed early in the 20th century, the first real steps toward space flight were taken after World War II when a survey of advancements in rocketry was taken. The V-2 weapon of World War II was large enough to carry payloads above the atmosphere. The advanced efforts of the German scientists at Peenemunde followed by rocketry development in the United States and the Soviet Union led to a growing acceptance of space as a promising area of exploration. But Russian and American rocket development took two separate roads. In the United States, the creation of thermonuclear weapons made it possible to construct lighter, and even more potent, warheads than the bomb that was dropped on Hiroshima. Plans for large and heavy missiles were scrapped and planning began on smaller rockets, more in tune with the sizes of warhead packages. On the other hand, the Russians concentrated on huge missiles because their thermonuclear technology wasn't going very well. But, of course, that development placed them in the leading role to enter space first. Until 1957, Western interest in space exploration was limited to small groups of scientists from select military research labs and British and American universities.

Then, quite unexpectedly on October 4, 1957, the space age was suddenly ushered in when the Russians launched the first Sputnik. Americans couldn't believe the Soviet Union had beaten the United States into space. What made matters even worse was the launching of a second, much heavier unit with a dog named Laika in it less than a month later.

A congressional inquiry into the U.S. space efforts took place. The pressure created by Sputnik and the congressional investigation may well have precipitated a speeding up in the U.S. space program. And, on December 6, 1957, only four seconds after liftoff, a first-stage propulsion failure destroyed the first full-scale test vehicle of America's Vanguard satellite program. That embarrassment led to stormy discussions in Congress and in the media on why the U.S. was lagging in the area of high technology, what the Russian lead in space really meant, and what could be done about regaining U.S. superiority in advanced science and engineering. In 1958, Congress passed legislation which President Eisenhower signed creating the National Aeronautics and Space Administration (NASA). NASA space goals were short and specific: emphasize scientific and technological developments, discoveries with significance for national security, knowledge gained in the military space effort by the civilian program, international cooperation, coordination of U.S. scientific and technical resources with respect to space, and wide dissemination of technical information with benefits to mankind.

But, to the chagrin of national pride, the Soviet Union had the exclusive on space spectaculars: three lunar probes and the first orbiting of a human being around the earth in 1961. Few were pacified during this time that the first two small U.S. satellites made discoveries of major scientific significance: the slight "pear shape" of the earth and the Van Allen radiation belt.

Apollo 16 commander John W. Young on the moon, April 21, 1972.

The 1962 to 1969 phase focused upon one major goal: land a man on the moon and bring him home after a short stay. Such a goal provided NASA a cohesive, goal-oriented plan which would capture the interest of the public and Congress. Spending for a manned space flight jumped from 25 percent to 75 percent of the total NASA budget. Soon the Soviets began to fall behind as the Mercury, Gemini, and Apollo programs provided one success after another, culminating in a lunar landing in July of 1969. But as successes mushroomed, so did public debate on the amount of money being spent on such efforts. There was particularly heavy criticism of space activity in 1967 when a fire during an Apollo test caused the deaths of three astronauts. But when the lunar landing was finally achieved, the magnitude of the success, its unprecedented nature, and the consequences for humanity were fully appreciated, the American public recognized the achievement as a profound accomplishment.

In early 1972, NASA was authorized by Congress and the president to proceed with the development of the space shuttle, which held the promise of lower transportation costs as well as making possible new concepts of space operations. As blueprints for a shuttle began pouring in, there began emerging a dramatic and constantly accelerating application of space-derived scientific and engineering knowledge. Suddenly, more possibilities and achievements were apparent: intensive analyses of the moon, planets, asteroids, sun, and stars; new breeds of Earth Resources Technology satellites and Skylabs and, all-new types of communication satellites.

Thus, if it hadn't been for the concentration during the '50s and '60s of technology, skill, accelerated experience, and high visibility which was both the cause and consequence of space efforts during the 1970s, this capability for America's phenomenal space successes would never have come about. Now, insists NASA, comes an even greater challenge for space activities—one that is broader than that which culminated in the landing on the moon.

In planning a long-range space program for the coming decades, scientists insist that they are responsive to mankind's needs and wants, while building a solid foundation of ethical responsibility and technological capability toward a future of increased options and choices. That alone will provide validity to a space program which is aiming for nothing less than these far future space endeavors:

- A multi-element Near-Earth-Orbit space community consisting of a number of separate settlements and facilities in various orbits around earth (e.g. space manufacturing plants, energy plants, apartments, recreation, tourism, hospital, sanatorium, research center, space vehicle hangars, propellant farms, agricultural centers, launch facilities, etc.);

- A monolithic space city housing various commercial facilities such as hotels, recreation areas, restaurants, hospitals, and residences;
- A lunar colony and mining base which has achieved virtual self-sufficiency and provides material, scientific data, and various services to earth;
- Solar power plants in space, orbiting nuclear reactor power plants, orbiting power plants, and the passive power relay satellite, all of which deliver energy to earth by microwave beams;
- A Mars settlement covered by a large dome to provide environmental enclosure and protection against cosmic radiation;
- Colonies and mining bases on or inside large asteriods;
- Manned exploration of outer planets, solar systems, stars, and universes;
- And, man-made worlds.

The merits of such far-future space endeavors lie in the fact that they can benefit all people on earth. As self-sufficiency in space is approached with its basically non-elitist philosophy, the true humanization of space begins.

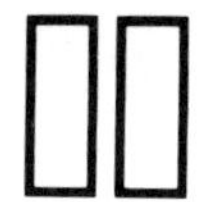

Man Invades Space: The Limitless Frontier

This decade will long be remembered for the bold initiatives and high scientific productivity of the planetary programs of the United States.

Dr. James C. Fletcher
NASA Administrator

A new era of space exploration and humanization will come into being during the early 1980s with the advent of the Space Shuttle and its ability to economically transport a variety of heavy payloads into orbit. As the name suggests, the shuttle would be capable of ferrying both people and equipment between earth and space. It has been designed to reduce the cost and increase the effectiveness of using space for commercial, scientific, and defense needs. NASA officials feel that with its versatility and reusability, the shuttle will truly open the door for the routine use of space. Its projected lifetime of up to 100 round trips would lower the cost of space flight substantially. Further savings would come from the shuttle's passenger-cargo payload capacity of up to 70,000 pounds. With its increased operational flexibility, the shuttle has become the pivot in the NASA pro-

The Space Shuttle is unveiled, September 17, 1976.

gram for the late 1970s. In giving the green light for shuttle development in 1972, President Richard Nixon predicted the vehicle would become the "workhorse for our whole space effort." Indeed, as a simple, economic transportation system to earth orbit, it will provide the workhorse capabilities of such earthbound transport facilities as airlines, ships, and trucks. Economists feel it will be as crucial to the nation's future in space as the more conventional carriers of today are to America's life and well-being. Dr. Fletcher, NASA's Chief Administrator, says, "We have found that the benefits made possible by Space Shuttle use in the twelve years from 1980 through 1991 will average more than $1 billion per year. Most of the benefits will result from the decreased cost of payloads and from their reuse, but the shuttle also offers significant savings in transportation, as well. . ."

A comprehensive NASA report, "Outlook for

Space,'' prepared by a special study team of space scientists and economists identified specific contributions the shuttle program could make in the next twenty-five years to meet the needs of people and nations. One such grouping of contributions, ''earth-oriented activities responsive to basic human needs,'' lists the shuttle's use of satellites and laboratories in space:

Food and forestry resources

- Global crop production forecasting
- Water availability forecasting
- Land use and environmental assessment
- Living marine resource assessment
- Timber inventory
- Rangeland assessment

The environment

- Large-scale weather forecasting
- Weather modification experiments support
- Climate prediction
- Stratospheric changes and effects
- Water quality monitoring
- Global marine weather forecasting

Life and property

- Local weather and severe storm
- Tropospheric pollutants monitoring
- Hazard forecasting from *in situ* measurements
- Communication-navigation
- Earthquake prediction
- Control of harmful insects

Energy and mineral exploration

- Solar power stations in space
- Power relay via satellites
- Hazardous waste disposal in space
- World geologic atlas

Transfer of information

- Domestic communications
- Intercontinental communications
- Personal communications

Scientific and commercial purposes

- Basic physics and chemistry
- Materials science
- Commercial inorganic processing
- Biological materials research and application
- Effects of gravity on terrestrial life
- Living and working in space
- Physiology and disease processes

Earth science

- Earth's magnetic field
- Crustal dynamics
- Ocean interior and dynamics
- Dynamics and energetics of lower atmosphere
- Structure, chemistry, and dynamics of the stratosphere/mesosphere
- Ionosphere-magnetosphere coupling

The Space Shuttle era will begin approximately twenty years after the first U.S. venture into space. Since that eventful Thursday morning in 1958, unmanned satellites have probed the near and distant reaches of space. Manned systems have been used to explore the lunar surface and expand the present knowledge of the earth, moon, and sun, as well as the adaptability of humans to extended space flight in near-earth orbit. With the Space Shuttle program, a new era of routine operations in space will take place. In-orbit maintenance of satellites in earth orbit and continuous, manned operation of space stations will become everyday practice. This will result in the construction of large and complex structures in space, and will lead to many more benefits to people on earth such as solar power satellites. Spacelabs will be carried aloft by the shuttle in support of manned orbital operations. Free-flying or automated satellites will be deployed and recovered from many types of orbits. Automated satellites with propulsive stages attached will be deployed from the shuttle and placed in high-energy trajectories. Such an approach to space operations will provide various channels for conducting explorations.

SPACE SHUTTLE MISSION AND SYSTEM COMPONENTS

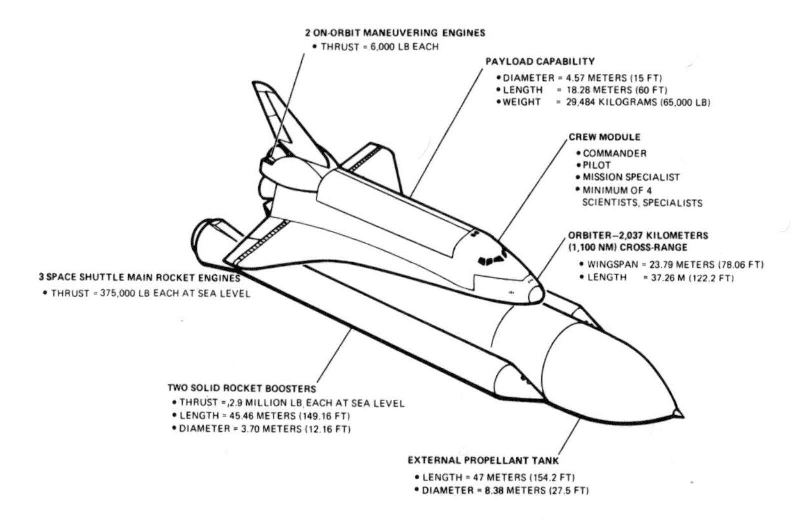

The Space Shuttle Transportation System consists of an Orbiter spacecraft, two solid rocket boosters for the launch, and an external tank (ET) to house the liquid hydrogen fuel and the liquid oxygen oxidizer for the necessary three 470,000-pound-thrust main engines. The Orbiter and solid rocket boosters (SRBs) are capable of repeated flights, while the external tank is expended on each launch.

The shuttle mission begins with the installation of the

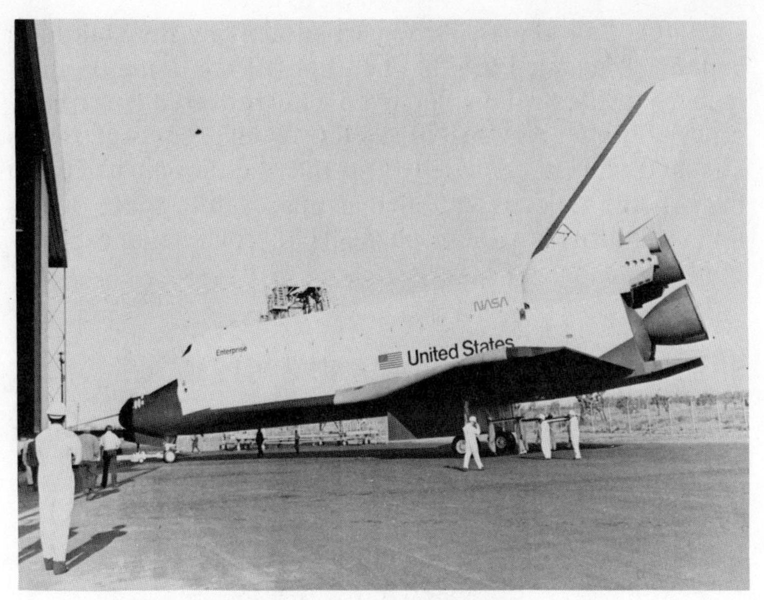

The first Space Shuttle.

space payload into the Orbiter's bay. Scientists and engineers will check and service each payload and will activate it in orbit. The Orbiter's main engine and SRBs will ignite and fire in parallel at liftoff. After burnout, the two SRBs are jettisoned and recovered by means of a parachute system. Before the Space Shuttle Orbiter reaches orbit, the large external tank is jettisoned. The Orbiter's orbital maneuvering system (OMS) is used to attain the desired orbit and to make any subsequent maneuvers that may be required during the mission. When the bay doors in the top of the Orbiter fuselage open to expose the payload, the crewmen are ready to begin payload operations.

At the conclusion of orbital operations, maneuvers for deorbiting are initiated. From a high angle of attack, reentry is made into the earth's atmosphere. Then, at a relatively low altitude, the Orbiter goes into horizontal flight for an aircraft-type approach and landing. The average duration of the mission will be seven days. However, the mission dura-

Liftoff of the Orbiter spacecraft, two solid rocket boosters and external liquid fuel tank.

Into orbit.

The solid rockets are jettisoned at about 27 miles altitude.

The external liquid fuel tank is jettisoned in orbit.

Space Shuttle Orbiter places earth resources satellite in orbit.

tion can be extended to as long as thirty days if the necessary equipment and adjustments are added to the spacecraft.

The Space Shuttle Orbiter is designed and constructed to lift into orbit a crew of seven, including both scientific and technical personnel, along with their payloads. The remainder of the vehicle system (SRBs and external fuel tank) is used solely to boost the Orbiter into space. During flight, the smaller Orbiter rocket engines of the orbital maneuvering system (OMS) provide control and maneuvering. While flight in atmospheric conditions is taking place, the Orbiter is controlled by the aerodynamic surfaces on the wings, as well as by the vertical stabilizer.

During a normal mission, the Orbiter will remain aloft for a week or so, return to earth with its personnel and payload, glide in for a landing like a commercial jet and be prepared for another flight within fourteen days. Not only this, but the shuttle will be readied for a rescue mission launch from standby status within twenty-four hours after an alert. For immediate emergency rescue, the cabin will take as many as twelve people. Thus, all the scientists and engineers of a crippled Orbiter could be rescued by another shuttle.

The SRBs, which burn along with the Orbiter main propulsion system, separate from the Orbiter and external tank at an altitude of 24 miles. They then descend on

Orbiter fires engines to change orbital path.

Orbiter returns to earth's atmosphere.

Pure silica tiles protect the Orbiter from the intense heat during re-entry.

Orbiter runway landing similar to that of a jet transport.

parachutes and land in the ocean approximately 175 miles from the launch site. There, they are retrieved by ships and returned to the launch site for reuse.

In the meantime, the Orbiter's main propulsion system continues to burn until the Orbiter achieves a velocity just below orbital requirements. The external tank then separates and falls into a remote area of the south Pacific or Indian Ocean, depending upon the launch site and mission. The OMS finishes insertion of the Orbiter into the planned orbit.

The passengers and crew are meanwhile sitting in a two-level cabin at the forward end of the Orbiter. The crew handles the launch, orbital maneuvering, atmospheric entry, and landing phases of the mission from the upper level flight deck. Payload handling is accomplished by crewmen in the aft cabin payload station. The area for seating passengers is on the lower deck. The cabin is open and spacious.

Mission flexibility is accomplished with a minimum of volume, weight, and complexity. Space flights will no longer be restricted to intensively trained, physically perfect astronauts. Now, the shuttle will accommodate experienced scientists and technicians.

Passengers and crew members will undergo a designed maximum gravity load of only 3g during launch and less than 1.5g during a normal reentry. Such accelerations are about one-third the levels experienced on previous manned flights. But a standard sea-level atmosphere will be the key feature with the spacecraft.

Although the shuttle will have the capability of conducting national and worldwide missions, the primary responsibility of the shuttle will be to deliver heavy payloads of 65,000 pounds into earth orbit. Also, payloads with propulsion stages can place satellites into high earth orbit or into lunar or planetary trajectories.

The shuttle is more than a transport vehicle. The Orbiter can conduct missions unique to the space program such as retrieving payloads from orbit for reuse, servicing or refurbishing satellites in space, as well as operating space laboratories in orbit. Such capabilities add to the productivity and flexibility of the missions.

Among the variety of uses Space Shuttle will have are a great number of applications to the environment of space and of space platforms. Such applications can be accomplished through operation of satellites, satellites with propulsion stages, space laboratories, or combinations necessary to specify missions. Also, the shuttle will offer a laboratory capability to conduct research and to explore techniques and equipment which may develop into new operational satellites.

But, according to NASA scientists, the shuttle will not be restricted to uses which can be forecast today. The reduction in the cost of earth-orbital operations and new op-

Orbiter flight control system.

Orbiter cabin work area with living quarters below.

Orbiter places satellite in orbit.

erational techniques will enable new and unforeseen solutions to problems.

Perhaps the most important shuttle missions will place satellites into orbit around the earth. The crew will consist of shuttle pilots, and mission and payload engineers. Once the spacecraft reaches the orbit selected, the mission and payload engineers will begin predeployment operations. When the satellite(s) is ready, the crew will signal the payload deployment system, which raises the satellites from the cargo bay retention structure, extending it away from the Orbiter, and releasing it. The final activation of the satellite will be handled by radio command. Then, the Orbiter will stand by until the satellite is functioning satisfactorily before proceeding with the rest of the mission.

The shuttle will also be capable of retrieving a satellite launched during an earlier mission. Such a satellite can be returned to earth for refurbishing and reuse. Five to seven small satellites can be delivered during a single mission. At

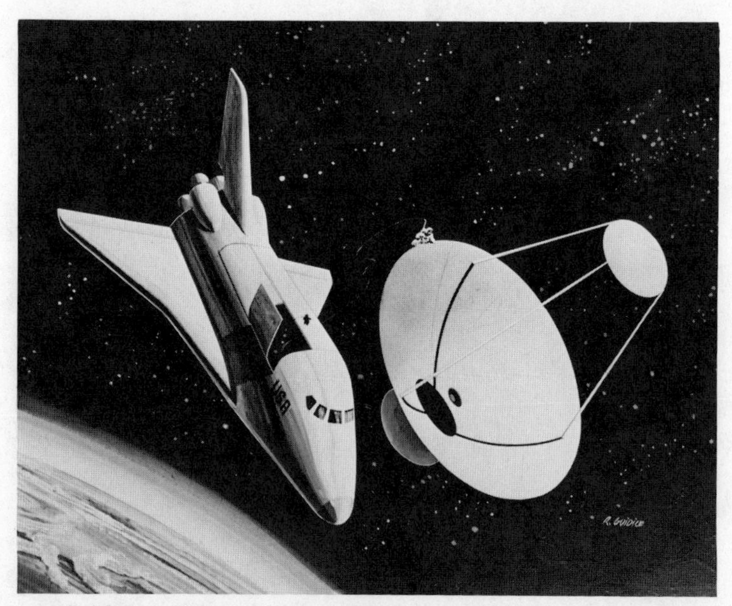

Orbiter will be able to perform maintenance on satellites in orbit.

Crew members operating from the Orbiter will assemble larger satellites and space stations.

the launch site, the satellites will be checked out, serviced, and placed aboard the Orbiter.

In order to retrieve a satellite for return to earth, the Orbiter will rendezvous with it, maneuver within a few yards of it, and clutch it with a remote manipulator arm. Once the satellite is deactivated by radio command, it will be inserted into the cargo bay and clamped into place. The Orbiter will conduct the deorbit maneuvers, reenter the atmosphere, and land.

The Space Shuttle Orbiter is engineered and constructed for a 14-day ground turnaround. From landing to relaunch, 165 hours of actual refurbishing will be necessary. As soon as the Orbiter returns from its mission in space, it has to undergo safing before payloads are removed and maintenance work undertaken. Such a safing operation includes draining and cleansing the propellant feedlines, as well as removal of explosive activators. The payload bay

Maintenance and refurbishment will take 14 days.

support equipment must be inspected and serviced. The thermal protection system, landing gear, main and auxiliary propulsion components, power systems, flight instrumentation units, and communication controls and networks also have to be analyzed and repaired. Such operations will require approximately three-fourths of the total processing time before relaunch takes place. From the maintenance hangar, the Orbiter will be towed to the assembly unit where it will be lifted vertically and erected on the mobile launcher platform. Then the mated shuttle will be towed to the launch pad to await countdown for another excursion into space.

For NASA officials and scientists, one of the most exciting benefits the shuttle offers is its ability to place free-flying scientific laboratories in space. For example, the space telescope represents an international facility for on-orbit space research controlled by scientists on earth. The shuttle would lift the telescope into orbit. Once there, crewmen would ready the facility for operation. During periodic revisits to the facility, the shuttle crew would inspect the various telescope subsystems, replace wornout parts and equipment and, some five years later, return the huge instrument and its facility to earth for a complete overhaul.

Another laboratory destined to be lifted into space by the shuttle is the long duration exposure facility (LDEF). The LDEF is a reusable, inexpensive, unmanned laboratory upon which a number of long range experiments can be mounted in order to determine the effects of exposure to space. After a year in orbit, the facility will be returned to earth for study.

However, the most activity is being predicted for geosynchronous orbits (an equatorial orbit with the satellite traveling at the same speed as the earth thereby remaining in the same spot relative to the earth's surface), deep-space missions, elliptical orbits and higher circular orbits. Payloads with such destinations necessitate a propulsion stage in addition to the shuttle. Like other payloads, the shuttle will lift the satellite and the propulsion stage into orbit. Before release there, the combined propulsion-stage/satellite unit will be inspected and prepared for launch. During this time, the guidance information will be revamped and updated. The Orbiter will then back a safe distance away before scientists on earth issue radio command signals to fire the propulsion state engines.

Both visual and remote monitoring can be done by the shuttle payload crew. In case of a malfunction, both the satellite and stage can be retrieved for analysis and repair. In case scientists discover that repairs are beyond the shuttle's onboard capability, the whole payload unit (satellite and propulsion stage) would be returned to earth for repair.

NASA scientists look forward to the launching of the Mariner Jupiter Orbiter by the shuttle during the early part of the 1980s. Such a launch would be the first. The purpose of the Mariner Orbiter is to gather additional data about Jupiter, its satellites, and the space surrounding it.

Currently, NASA scientists and engineers at the Goddard Space Flight Center are creating a new family of modular spacecraft satellites to be inserted into orbits at various inclinations and altitudes. Inexpensive and already-standard hardware is expected to make up each satellite. A key feature of the satellites is that they can be serviced by changing the supporting minor system units and application assemblies. This feature coupled with the shuttle's attached equipment and operational maneuvers will allow maintenance and replacement of the satellites while in orbit. Added to the great weight and volume capacity of the shuttle, this added capability offers the payload engineers latitude to create and control satellites which can lower payload costs as well as enhance performance.

But, perhaps, the most promising of all the shuttle's features is Spacelab. Says NASA's Dr. Fletcher, "This is an unprecedented cooperative enterprise which represents a most generous contribution by the European nations to the basic space facility of the 1980s, one which we can use in common on either a cooperative or reimbursable basis."

Originally developed by the European Space Agency (ESA), Spacelab is truly an international program. The big pressurized module with its extensive external equipment pallet will be a frequent payload vehicle during this new era in space. What excites scientists most is that Spacelab will offer an extension of the experimenter's ground-based labs with additional features such as long-term gravity-free environment, a location from which earth can be observed and analyzed as an entire planet. Thus, this lab will serve as a "celestial city," a small habitat which will allow earth to be examined free of atmospheric interference.

Three or four Spacelab habitats will be launched. One includes a pressurized module where scientists can experiment in a room temperature, shirt-sleeve environment. A tunnel will connect the Orbiter crew compartment with the Spacelab. Technical instruments will be mounted on a pallet aft of the pressurized module if they need exposure to the space vacuum or are too large and bulky to fit inside. The Orbiter will circle in an inverted position in order to orient all the measuring and other technical instruments toward earth for analysis of the planet's resources, as well as for investigations of geophysical and environmental characteristics.

Several other Spacelab habitats include those which have a large pallet upon which numerous instruments are installed and controlled from the payload engineer's station within the Orbiter. These will serve in place of pressurized modules. But obviously, pressure-suit operations in the payload bay will be necessary when instrument service is needed.

Ten European nations with strong interest in Spacelab experimental habitats have committed themselves to spend $500 million to design, construct, and deliver one flight unit to the United States. Contracts stipulate purchase of additional units by the United States. The ten nations in the joint venture include Austria, Switzerland, Denmark, Netherlands, Spain, Belgium, the United Kingdom, France, Italy, and West Germany. The international brigade of space scientists from these nations feel that many types of scientific, technological, medical investigations, as well as applications, can be created with the newly designed flight hardware. Since each Spacelab will be flown as often as fifty times during its projected normal lifespan of one decade, this will offer a brand new capability for manned participation. This will obviously increase the effectiveness of space research as well as reduce the cost of space technology. The crew teams and scientific payloads will be international in origin, although many will be provided by the United States.

HISTORY AND CRITICISM OF THE SPACE SHUTTLE

Despite all the earthly benefits of the shuttle, the program has its detractors. Many critics felt when the concept was proposed in 1972 that the project would cost many times more than estimated. Although NASA felt the Space Shuttle was the ideal successor to Apollo, a program then drawing to a close, and although President Nixon endorsed construction of the reusable space vehicle, critics insisted there was no present or foreseeable need for a spacecraft that would have to make one flight a week to justify its development costs. Some of those objecting to the tremendous costs believed that the project was nothing more than a welfare program for the crumbling aerospace industries. They demanded that shuttle monies be spent on the poor, jobs, health care, and rapid transit.

On Sunday, March 7, 1970, President Nixon announced two of the nation's "relatively modest goals for the latter part of the decade." He said, "By no means should we allow our space program to stagnate. But with the entire future and the entire universe before us, we should not try to do everything at once: our approach to space must continue to be bold—but it must also be balanced." Nixon asserted and most NASA officials agreed, the nation's space program had to be guided by three general principles: exploration, scientific knowledge, and practical applications for earth. To fulfill these goals, he listed six specific objectives: (1) continuing exploration of the moon; (2) bolder explorations of the planets and the universe; (3) substantially reducing the cost of space operations; (4) extending man's ability to live and work in space; (5) expanding to American industry the practical applications of space technology; and (6) encouraging greater international cooperation in space.

But to space agency officials, the "bold" yet "balanced'' program was nothing but an austerity program. A few months earlier in October of 1969, a cabinet-level Space Task Group headed by Vice President Spiro T. Agnew had proposed a much broader and specific set of goals for the '70s and '80s. Surprisingly, the major recommendation of the Task Group report, entitled ''The Post-Apollo Program: Directions For the Future,'' was that the nation ''accept the long-range option or goal of manned planetary exploration with a manned Mars expedition before the end of the century as the first target.''

The recommendations also included three alternative programs designed to land men on Mars in 1983, 1986, and 1992. Regardless of which option was chosen, the Task Group felt four projects would have to be concluded in preparation for manned flight to the red, mysterious planet: (1) a manned space colony in permanent earth orbit; (2) a reusable space vehicle to carry men and equipment from earth to colony, or space station, and back; (3) a nuclear rocket for space flights of several years duration; (4) a nuclear-powered tug to move orbiting space stations and space vehicles.

Upon reading the recommendations, Nixon was at first noncommittal. A while later, however, the President agreed with the goal of landing men on Mars before the year 2000, but wasn't enthusiastic about the expense involved. In a nationwide address on March 7, 1970, Nixon explained, ''We will eventually send men to explore the planet Mars.'' And, he added, ''We must build on the successes of the past, always reaching out for new achievements. But we must also recognize that many critical problems here on this planet make high-priority demands on our attention and our resources.''

But nothing was done for almost two years. Then, abruptly, January 7, 1972, the President declared that the nation ''should proceed at once with the development of a reusable space shuttle that would take the astronomical costs out of astronautics.'' Such a spacecraft, Nixon felt, would "help transform the space frontier of the 1970s into familiar territory, easily accessible for human endeavor in the 1980s and 1990s." The approximate expenditure for designing and constructing two test flight vehicles over a seven-year period was estimated at $5.7 billion, approximately one-fourth of the Apollo moon program cost, and, would create at least 50,000 new jobs in the flagging aerospace industry.

But, of deep concern to Dr. James Fletcher, other NASA officials, and others in the aerospace field, was the rising animosity in Congress and among the general public toward expensive space technological innovation of all kinds. The full expression of that feeling arrived in 1971 when the Senate twice refused further development of the supersonic transport (SST) jet.

None other than Senator Walter F. Mondale, then a Democratic senator from Minnesota, led the opposition in Congress to the shuttle project. He had served notice that he would wage an all-out campaign to defeat it. Describing the program as a "senseless extravaganza in space," Mondale said that "in the magnitude of its cost, in the folly of its concept and, thus, in its damage to the country, the Space Shuttle is many times worse than the SST. The money proposed for shuttle development should go to solve human problems such as mass transit, housing, education, and the environment."

While serving in Congress, Mondale worked hard to delete funds for the creation of a shuttle from the fiscal 1972 NASA authorization bill. In a 1971 Senate speech, he argued,

> I ask the Senate, how are we going to sell to the American people such an expensive program? Has anyone in the Senate ever been approached by a constituent saying he needed a Space Shuttle or a space station? Do they think that a Space Shuttle and space station are as important as the problems that we face

> here on earth, such as the problems of environmental pollution, the decaying city, health care, and health research, and the other problems which face us here in our own society? If priorities mean anything at all, if there is any rational allocation of this nation's resources, then certainly we cannot continue to waste money on this program which lacks the support of any rational cost study.

Needless to say, most congressmen, as well as scientists, did not agree with Mondale. To them, the case for the shuttle rested on the assertion that it was needed to guarantee American technological supremacy in space. Equally adamant, Senator James B. Allen, a Democrat from Alabama, said during the debate, "Surely, this country cannot so soon have forgotten the tremendous shock we faced in 1957 when the first Sputnik went beeping around the world to give us warning that another country was challenging our position of leadership. It took us many years of hard work and unflagging effort and much of our treasure to regain that position of leadership, and today [1971] we see it again being strongly challenged."

The same feeling was echoed by numerous other congressmen who insisted America must not turn its back on technological progress. If the nation did, it would mean condemning a majority of Americans to progressively inferior standards of living and to become a "second-rate technological power, subject to coercion by other nations which are more power-conscious."

But Congressional opponents of the Space Shuttle did all they could to kill the project in its infancy. They questioned the purported launch-cost savings, raised doubts about the need for the shuttle payload capacity, and openly wondered whether unmanned exploration flights would be better than manned information-seeking flights.

Scientists, however, took the opposite view. For instance, Lyman Spitzer, Jr., a Princeton University astronomer, argued that:

> The role of man in connection with a large astronomical observatory in space will be most crucial. Manned missions are needed to perform maintenance and repair and to replace obsolete instruments with newly developed ones. Indeed, one may visualize a small team of men visiting such a telescope at intervals, replacing faulty modules, inserting improved instrumentation and checking the operation before returning to earth. The Space Shuttle is ideally suited to transport men on such space missions.

Thus, in 1972, the Space Shuttle controversy raged throughout the nation. The AFL-CIO Executive Council felt the shuttle would aid the science and defense industries, as well as provide 50,000 aerospace jobs, primarily along the west coast states. Opponents found such arguments unconvincing. It seemed that the fate of America's manned space flight program was hanging in the balance. With only two more Apollo moon missions scheduled, the National Aeronautics and Space Administration was finding itself without a new long-range mandate to send men into space. However, the agency did not have to worry too long. In endorsing the Space Shuttle program, President Nixon said, "The shuttle project involves, as do many other technological programs, the question of national priorities. Its detractors foresee only limited benefits, at best, sometime far in the future. Its supporters claim, and I agree with them, it will help stimulate the economy in the short run and lead to long-term improvements in the quality of life on earth."

The undeniable fact was that the program would help to create jobs in a time of high unemployment—a most persuasive argument during a presidential election year.

SHUTTLE RESERVATIONS NOW BEING ACCEPTED

The Space Shuttle wasn't even flying on its own yet, when NASA began selling cargo space aboard the Orbiter in the

The Orbiter *Enterprise* mated to a Boeing 747 during landing tests.

be turned over for military use with launches from California's Vandenberg Air Force Base.

Costing approximately $6.9 billion, the shuttle program is being developed under NASA's specific direction. The agency's Lyndon B. Johnson Space Center in Houston is responsible for integration of the complete system and for the development of the Orbiter, while NASA's Marshall Space Flight Center in Huntsville is responsible for the development of the shuttle main engines, the external tank, the solid rocket boosters, and a number of scientific and applications payloads. Launch, landing, and turnaround operations will be handled by the Kennedy Space Center.

One of the most fascinating technological achievements to occur during the development of the shuttle dealt with the creation of a thermal protection system for the craft. LI-900 (which stands for Lockheed Insulation/9

spring of 1977. Making up cargo manifests for th
flights in 1980, Chester Lee, Director of Space Trans
tion Systems for NASA, said, "Anybody and any
can buy cargo space as long as the payload is safe
good taste. The rates will only be about 60 percent
pendable vehicle launch costs. They can buy the
cargo space or just a portion of it. The cargo space av
is 60 feet long, 15 feet in diameter and can take
pounds. Those who don't need all the cargo hold wil
portion of the cost."

According to Lee, there are three separat
policies. For instance, the U.S. Department of D
will receive the bargain rate of $12.2 million for
cargo bay. On the other hand, civilian U.S. gove
agencies will pay $16 to $19 million for the same
Commercial and foreign groups will be charged $1
million for the same service.

Explained Lee, "We're already negotiating c
with the COMSAT Corporation, TELESAT of
and Satellite Business Systems to carry satellites in
in 1981. Each one of these companies must put up $
earnest money."

Although some critics have dubbed it the era o
tation, rather than exploration, the new era in sp
under way with shuttle glide tests at the Dryde
Research Center near Edwards Air Force Bas
Mojave desert. Dubbed the "Enterprise" after th
ship in the "Star Trek" television series, the shu
flew aloft piggyback-style on a Boeing 747. Before
of 1978 the Orbiter will be sent to Huntsville, Alab
further design.

Right on schedule, orbit tests will begin late
from the Kennedy Space Center. Around the su
1980, three civilian shuttles are due to begin round
erational flights hauling communication satell
tronomical equipment, earth resource sensors, a
laboratories into orbit. Two additional 123-foot shu

pounds-per-cubic-foot) is a high purity silica fiber insulation which will protect 70 percent of the shuttle's exterior against temperatures up to 2300°F created when it reenters the earth's atmosphere. Each shuttle vehicle will require approximately 34,000 "tiles" of LI-900 in thousands of different sizes, shapes, and thickness. The space-age insulation, which sheds heat so efficiently it can be held with bare hands while still red hot, can withstand repeated heating and cooling, for up to 100 shuttle flights without replacement. Its resistance to thermal shock is unmatched in the history of engineering technology. The material can be taken from a 2300°F kiln and immediately immersed in cold water without damage. Surface heat dissipates rapidly and heat transfer from inside is so slow an uncoated tile can be held bare-handed by its edges seconds after removal from the oven. The basic raw material for the LI-900 is a common sand. The Lockheed Missiles and Space Company at Sunnyvale, California, will be manufacturing the tiles at the rate of 5,000 per month. The company has more than fifteen years experience with the material, and selected it for the

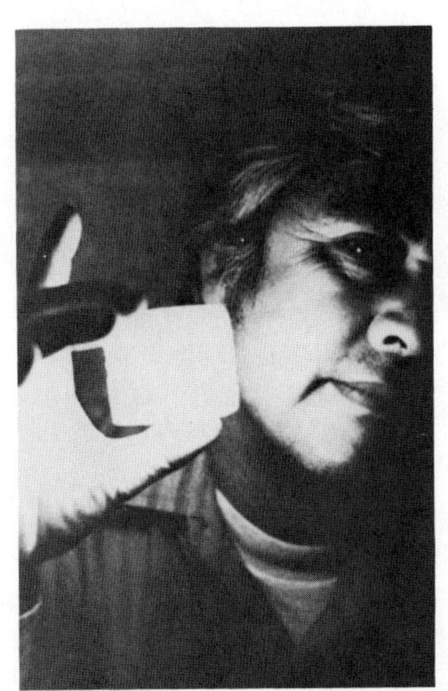

Silica brick heated to 2300°F held by a technician only 10 seconds after removal from the oven.

shuttle because of its lightweight, low thermal conductivity, low thermal expansion, and high-temperature stability.

Unlike the heat shields of the Apollo moon spacecraft which survived only one mission, tests on nine tiles of insulation conducted in mid-1977 indicated that after a full month of withstanding the highest temperatures and pressures man can create the tiles remained unchanged.

When the Space Shuttle becomes operational in 1980, it will be an important tool, and provide mankind with information to help in managing and preserving the crowded earth. Users of the versatile shuttle system will include communication networks, research foundations, universities, observatories, federal departments and agencies, state agencies, county and city planners, public utilities, farm cooperatives, the medical profession, fishing industry, transportation industry, and power generation as well as water conservation planners.

Thus, payloads launched by the shuttle will provide practical data which will affect both the daily lives of people and the long-term future of mankind.

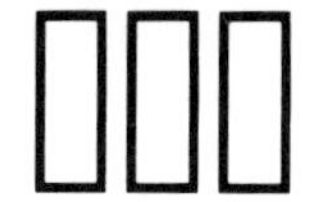

The Permanent Occupancy of Space: Earthly Payoffs Tomorrow

> It should not be surprising that we present no new ideas about what might be done in space. The world abounds in imaginative people, and the catalog of what might be done in space is immense. What we have done is to identify some of those possibilities which are within reach over the next twenty-five years and, most importantly, can contribute in a significant way to the needs of peoples and nations.
>
> Donald P. Hearth,
> Study Director, NASA

By the end of the 1960s, man had walked on the moon for the first time. In the 1970s, we sent our scientific sensors exploring across incredibly vast interplanetary expanses. After the launch of Space Shuttle in 1980, we will begin to move and inhabit space for good.

"The permanent occupancy of space," says Captain Robert F. Freitag, Deputy Director of Advanced Programs at NASA's Office of Space Flight, "is indeed our major goal for the 1980s." Speaking for NASA officials and scientists, Freitag echoes NASA's objective that the main thrust of future activities centers on establishing permanent orbital facilities engaged in projects of immediate and large-scale payoffs to earth.

With permanent occupancy and activities designed to exploit zero gravity and other unique environmental conditions, everyone can enjoy benefits—most of which are only vaguely understood and perceived—through the industrialization and commercialization of space. During the mid-1980s, the goal is for men and women to inhabit and work in space, establishing new industries, creating new materials and products, and initiating various scientific projects. Says Freitag, "All of this would be aimed at doing and providing for humans what they cannot do or make, at all or as well, on earth."

And now, at a time when the concept of providing easy access to space at reasonable cost via a cargo-carrying, reusable Space Shuttle is about to be implemented, many large-scale orbital facilities are being planned. Because of Orbiter's versatility in space construction, such projects as public service platforms, space "mini-factories," and satellite power systems are being designed for liftoff during the 1980s. Of all these concepts, perhaps the most important for earthlings concerns the solar power satellites.

SOLAR POWER SATELLITES

In light of our rapidly depleting supplies of fossil fuels, we must develop new sources of energy. It is the essence of the energy crisis. Space scientists feel the answer will be found in the development of a non-depletable source of energy which can be transformed into usable electricity. And, earth's primal energy source, the sun, provides an excellent solution. By harnessing its rays in space, where of course they exist without night, and transforming this power into electricity usable on earth, we could tap a power source more enduring than those we are depleting on earth.

Such a feat could be accomplished through the use of a

solar power satellite, a spacecraft the size of a small city which would be able to produce twice the usable power generated by Grand Coulee, the nation's largest hydroelectric dam. Forty-five of these satellites would be able to equal the total present electrical generating power of the United States. This would free oil, coal, and their derivatives for other important needs.

Stationed some 21,000 miles (35,000 kilometers) above the equator in geosynchronous orbit, these bright satellites would appear stationary when seen from earth. They would be bathed in sunlight 99 percent of the time, passing through earth's shadow only for very short periods in the fall and spring when it is late at night on earth. The satellites would catch and transform the sun's rays into electricity which would then be converted to microwaves beamed to large antennae on earth. The huge antennae would reconvert the power into simple electricity for America's power grid.

Since 1972, the Boeing Aerospace Company in Seattle has been the chief investigator of solar power satellites, using both its own research funds and contract awards from NASA and other governmental agencies. Boeing studies have focused upon two basic configurations for the satellite design: photovoltaic and Brayton heat engine. Each would be capable of creating 12,000 megawatts of usable power, enough to light and heat the needs of over a million homes.

The photovoltaic satellite would be rectangular in shape and cover an area of about 15.5 miles (24.9 kilometers) by 3.2 miles (5.2 kilometers), an area of almost 50 square miles (129 square kilometers), or the size of a small city. On this vast platform would be mounted about 14 billion solar cells. Such a spacecraft would have a mass of approximately 89,000 to 112,000 tons. The satellite's solar cells would transform sunlight directly into electrical energy, in the same manner solar cells power small satellites.

On the other hand, the Brayton heat engine satellite would use a series of four parabolic dishes, each about 3.5

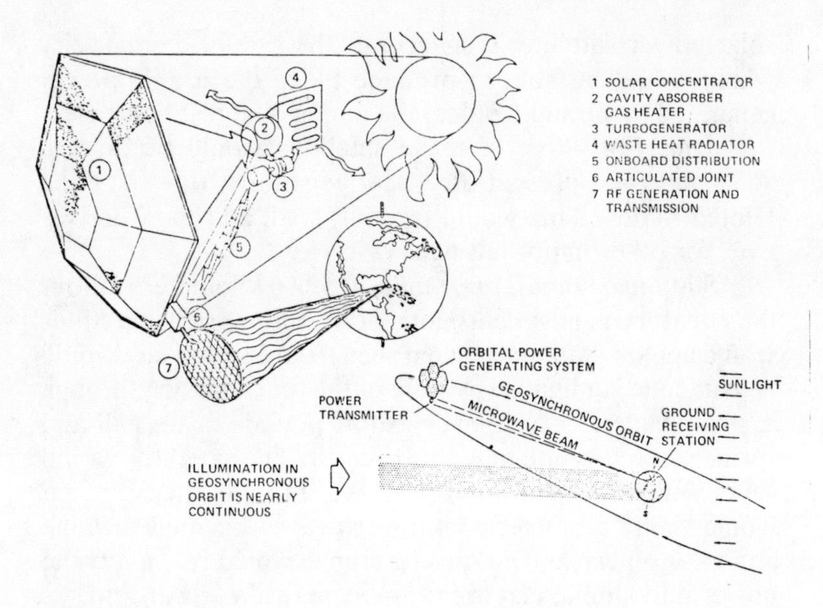

miles (5.6 kilometers) across. Together, they would be made up of thousands of steerable, extremely thin plastic reflectors. Such reflectors would direct the sun's rays into a domelike cavity absorber (a solar furnace) located over each dish. The concentrated sunlight would superheat gases that would expand and drive a series of turbo-generators girdling the absorber. These generators would create the satellite's electricity. Once through the generators, the heated gases would flow through pipes into large fin-like radiator panels in which the gases would cool before being recirculated into the cavity absorber to begin a new cycle. Such a heat engine satellite would weigh about the same as the photovoltaic satellite, some 89,000 to 112,000 tons.

The photovoltaic system is the less complex system. However, solar cells at the current stage of development are less efficient than thermal cycle engines and are more expensive since they require advanced manufacturing techniques. On the other hand, Brayton heat engine satellites are complex systems using complex devices. However, they have a higher conversion efficiency, are able to process large quantities of power, and have already been proved through the large-scale production of energy on earth. Although each system has definite attractions and drawbacks, space scientists feel both satellites appear to be very feasible.

And, getting the energy from space to home will be a relatively easy matter. Solar power satellites, whether heat cycle or solar cell, produce direct current electricity. This would be converted to microwave energy and channeled through two transmitters on each satellite. These antennae would be a little more than a half-mile across (one kilometer). Except for size,they would be an extension of current radar technology. These long antennae would direct the microwave beam to ground receiving antennae that would rectify the microwave energy to direct current electricity and be fed directly into the nation's power lines.

Rectenna installation, five by eight miles, to receive microwave transmission from orbiting solar power stations.

The rectennae (rectifying antennae) would measure about five miles (8 kilometers) by eight miles (13 kilometers) and would resemble cyclone fencing mounted in strips high enough off the ground for the area beneath it to be used for animal grazing or farming. Microwave levels outside and even beneath the antenna area would be below the already-stringent standards now in use. The beam itself would be of a low enough intensity to actually permit birds and other forms of life to pass through it without harm. The peak intensity at the beam center would be far below lethal levels even at long exposure. It would have no effect on aircraft or their crews and passengers.

Of course, satellites measured in miles rather than feet cannot be launched like the simple Apollo mooncraft. Therefore, solar power satellites would be built in low earth orbit for later shipment to higher geosynchronous orbit, or would be constructed directly at the higher orbit. In either

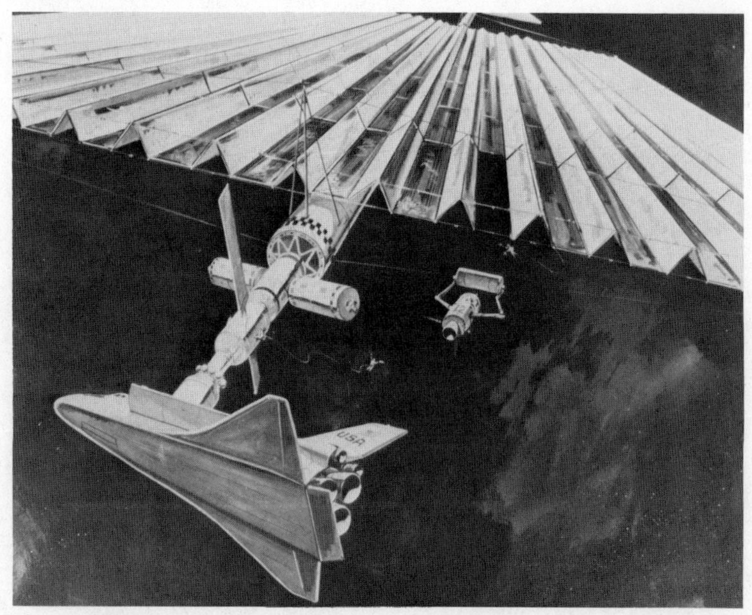

An artist's conception of a solar power conversion system in geosynchronous orbit.

case, the creation of objects in space the size of small cities is entirely outside human experience. Therefore, a whole new transportation system will have to be built.

For solar power satellites, large unmanned freighters, known as Heavy-lift Launch Vehicles, would carry outsized cargo pallets into low earth orbit where these pallets would be deposited and directed to docking stations at a space construction base. Like the shuttle, these huge freighters would be entirely reusable and available for relaunch within a short period of time.

NASA and Boeing space scientists say that today's Space Shuttle Orbiter with minor modifications will be the backbone of the manned transportation system, carrying the several hundred workers needed on the orbiting construction base.

Currently, Boeing engineers are designing orbital transfer vehicles to carry workers and equipment from

near-earth-orbit to geosynchronous orbit, as well as propulsion units to carry satellites and their units to a stationary position. In all likelihood, we may send into orbit as much as one million tons of hardware each year.

Such expanded utilization of space and its transportation systems will lower costs. For example, payload transportation costs for Vanguard in the late 1950s ran $500,000 a pound for payloads in the 20–30 pound range. For the Thor rocket during the 1960s, the cost was reduced to $10,000 a pound for payloads of little more than 1,000 pounds. Saturn moonrockets during the late 1960s and early 1970s cost $600 a pound for quarter-million-pound payloads. During the Space Shuttle era, NASA projects $150-a-pound payload costs attained through Orbiter reuse. Actually, by 1985, say industry studies, a payload cost for heavy lift launch vehicles will range in the neighborhood of $10 to $15 a pound. And, engineers predict the day will come during the decade after 1985 when huge space freighters will be leaving space centers at the rate of five to ten a day.

Boeing and other manufacturers are presently attempting to determine the costs involved for solar power satellite construction and use. Obviously, they must be built and operated at a cost which will allow their expense to be passed on to the users of the electricity they furnish. A little math illustrates that the revenue from one solar power satellite producing 10,000 megawatts of electricity sold at a rate of 30 mills per kilowatt hour would produce $78.8 billion in thirty years. Forty-five satellites would thus produce more than $3.5 trillion. And, 30 mills is currently the cost of electricity generated by a new oil-burning generation plant.

SPACE TELESCOPE PROJECT

Space scientists have long since known that the atmospheric envelope which encases the earth significantly degrades astronomical observation even during the clearest conditions. In attempting to gaze through it, the astronomer finds sight distorted and dimmed.

With the shuttle carrying it into orbit, an eight-foot space telescope will orbit some 300 miles (500 km) above the earth's surface. Commissioned by NASA, Boeing has been constructing such a telescope.

From the 300-mile orbit, the telescope will be able to search deep into space, possibly even to the outer fringes of the universe. Thus, without an atmospheric veil, wavelengths in the ultraviolet and infrared portions of the spectrum will be observable. But far more important, scientists will be seeing with far greater clarity than ever before. Quasars, galaxies, and other celestial objects as much as seventy-five times fainter than those seen by the finest earthly telescopes will at last be within the space telescope's range. Celestial bodies will be observed with ten times better resolution than that attained by the largest

Full-scale mockup of the space telescope.

The telescope as it will look in the cargo bay of the Orbiter.

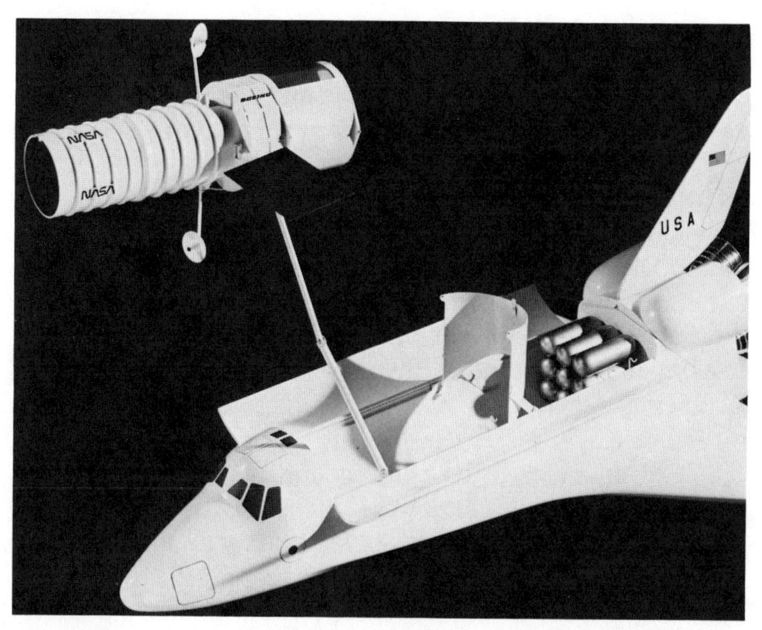

Deployment of the space telescope by the Orbiter.

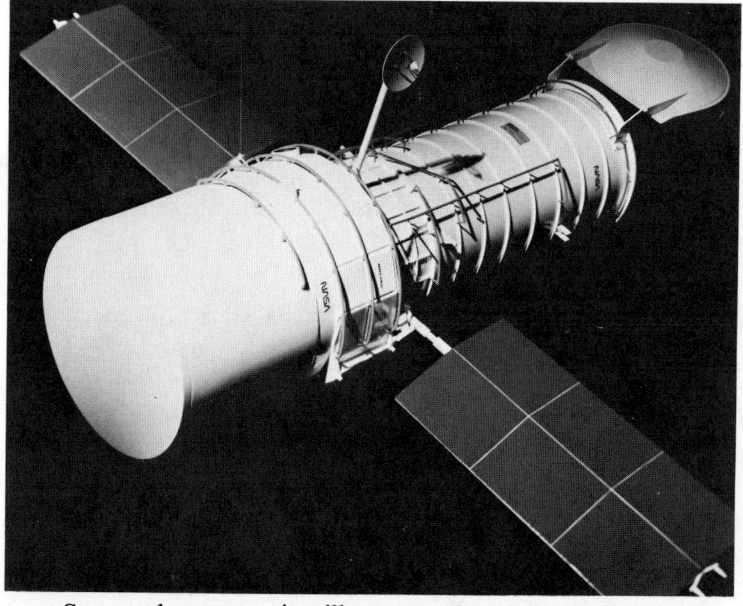

Space telescope as it will appear when fully deployed.

earth-based telescopes under the most favorable weather conditions. And, within this solar system, astronomers will be able to better monitor atmospheric and surface phenomena of the planets.

The space telescope will be carried into its earth orbit by the Space Shuttle Orbiter. Once in orbit, it can function unmanned for some twenty years. Astronauts will return to the massive telescope at various times in order to make in-orbit repairs and adjustments or to replace scientific instruments. Once every five years, the telescope will be returned to earth for refurbishment.

The diameter of its primary mirror will be about 2.4 meters, and the telescope will weigh approximately 15,000 lbs. Its overall length will be 45 feet with a total diameter of 15 feet. Electrical power for the telescope will be provided by solar panels when the telescope is on the sun side of its orbit and by batteries while it is on the dark side. Images detected by the telescope's science instruments will be transmitted periodically to the earth.

THE APPLICATIONS EXPLORER MISSIONS (AEM) BASE MODULES

Among the satellites which the Space Shuttle will carry and place into orbit are two small low-cost spacecrafts designated Applications Explorer Mission A (AEM A) and Applications Explorer Mission B (AEM B) which are being designed to explore regions of the earth and its atmosphere.

AEM A will have a Heat Capacity Massing Mission (HCMM), and AEM B is a Stratospheric Aerosol Gas Experiment (SAGE). In order to meet low-cost objectives, both satellites are based on a modular concept: an instrument module containing the mission's scientific instrument and its mission-unique parts, and a base module which will contain all necessary data handling, power, communications, command, and altitude control subsystems needed to support the science payload. The Boeing Aerospace Company's Space Systems Division has been designing, constructing, and testing these base modules at the fixed-price incentive contract of $8,236,000. The general contract is under the technical guidance of the Goddard Space Flight Center in Greenbelt, Maryland.

The base modules will be hexagons, approximately 28 inches (71 centimeters) across and 25 inches (64 cm) long weighing approximately 200 pounds (91 kilograms). Each module will contain solar panels with a total area of 24 square feet (2.2 square meters). The array on AEM A will be in a fixed position while that in AEM B will be stepped periodically to new positions. Both base modules will contain all the spacecraft's support instrumentation, including receivers, transmitters, power system, and altitude control systems. They will also serve as the platform for each spacecraft's scientific instrument module.

Placed into a 373-mile (600-kilometer) circular, sun-synchronous orbit by the shuttle, the satellite will be equipped with a radiometer with which it will sense the earth's surface temperatures at the hottest and coolest times of the day. This will provide data which will allow the determination of thermal inertia, or heat capacity, of various segments of the surface. Certain thermal inertias are characteristics of various surface materials and effects, as well as underlying substances. Accurate mapping of these temperatures by satellite would lead to benefits such as identification of rock types and mineral resource locations, measurements of foliage temperatures to determine the transpiration of water, an early indicator of plant stress. It would also assist in measurement of soil moisture effects by observing the temperature cycles of soils, mapping of thermal flows both natural and man-made, and prediction of water runoff from snowfields.

The AEM B satellite will accurately measure the decrease in solar radiation as it passes through the earth's atmospheric layers. As the satellite emerges from the earth's shadow during each orbit, a photometric device will absorb light from the sun and measure solar radiation. As the satellite continues in orbit, the line of sight from the satellite to the sun will scan the earth's atmosphere from the horizon up, resulting in a measurement of solar intensity at different layers. This procedure will be repeated in reverse during the satellite's encounter with the sunset. Each procedure normally requires about five minutes of operation.

During one year's operation, commencing with shuttle delivery in 1980, each of the two satellites will make about 10,000 measurements in the area between 72 degrees south and 72 degrees north latitude. By combining these measurements with data from the ground, scientists will be able to establish baseline data on global concentrations of aerosol and ozone and better understand the effect of transient phenomena on these concentrations. Scientists will also be able to gain insights into hemispheric differences, on the transport phenomena between atmospheric layers and on tropospheric/stratospheric exchange, and learn the effects aerosols and ozone have on the availability of solar energy on earth.

THE SOLAR SAIL DEVELOPMENT PROGRAM

Solar sail will bring to fruition an idea that has intrigued the imaginations of space scientists and engineers for almost fifty years: using the sun's photon energy to propel a large reflective sail on a free ride through space. The sail would employ a mirror-like aluminized plastic surface to attract the radiating photons which carry momentum. When reflected against this sail, the photons change momentum and a force is exerted against the reflective surface, much like wind against a sail. The speed of the solar sailcraft depends on its distance from the sun and the size, weight, and angle of the sail. The greater the sail surface and its proximity to the sun, the greater the reflectivity pressure or energy

Solar sailcraft.

thrust of the photons against the sail. The sail, the advocates argue, has the potential of vast improvement over ballistic (rocket) trajectories. Since it would carry no fuel, it would be cheaper than conventional spacecraft systems. By tacking against (or with) the solar photon stream, the solar sailcraft could fly toward the sun or away from it. NASA, and the Jet Propulsion Laboratory at the California Institute of Technology, will demonstrate the solar sail with the 1980 launch from the Space Shuttle toward the sun and outward to intercept Halley's Comet in March of 1986. For the first flight, the furled sail would be taken to the shuttle platform and erected by astronauts in the space vacuum.

The solar sail development program includes design of an 800-meter square plastic film sheet (half a mile on each side) only 2.5 microns (0.1 mil) thick, plus ultra-lightweight extendable booms for the spars and masts of the sail.

In addition to the square plastic sail, the heliogyro sail is another configuration under study at the Jet Propulsion Laboratory for solar sail missions. The name "heliogyro" is derived from the Greek word *helio* for sun, and *gyro,* for gyroscope which is the basic concept of the vehicle, a vehicle spinning in space, much like a gyroscope.

The heliogyro sail provides a huge reflective square to create a sail through use of twelve long, thin sail blades similar to the blades of a helicopter. These are used to reflect solar pressure as well as to control the vehicle. The plan is to allow the vehicle to spin and make use of centrifugal force to support and stiffen each blade. The spin of the sail will also aid deployment of the blades.

The sail blades can be pitch-controlled just like those of a helicopter to provide altitude control and maneuver the craft with respect to the sun. This will allow the thrust of solar pressure to either accelerate or slow the sail vehicle

Heliogyro sailcraft.

down. Increasing the velocity of an object in orbit around the sun allows it to overcome the sun's gravitational field and move outward. Reduced velocity allows the sun to pull the object inward. Thus, the spacecraft can either fly outward ito the solar system or inward toward the sun.

Once the heliogyro is lifted into earth orbit in a furled mode by Space Shuttle, it will be injected on an earth escape trajectory with an interim upper stage (IUS) rocket. The IUS will also be used to initially spin the vehicle, and centrifugal force will act to begin the extension of the blades stored on rollers. After the sail blades unroll to a length of 505 feet (154 meters), the no-longer needed IUS will be jettisoned. The blades will be sloped like those of a propeller to provide solar torque, which will continue spinning the vehicle. It will take two weeks for the sail blades to unfurl to full length.

When extended to its final length, each sail blade will be 3.9 miles (6.25 kilometers) long, 25 feet (8 meters) wide, and have 600,000 square meters of surface area. There will be twelve sail blades projecting from a central hub in two levels of six blades each. Each sail blade is made of .1 mil plastic material. An aluminum coating on the front of the sail blades would release heat and allow the vehicle to work close to the sun. Each blade would have edge reinforcements, and battens across the blade to provide stiffness. The rotational period of the fully deployed heliogyro is 200 seconds.

The advantage of the heliogyro sail over the square sail, argue many space scientists, is that the design's use of centrifugal spin provides blade support, reduces the need for a stiffening structure, and allows a lighter, higher performing spacecraft. The unrolling deployment of the heliogyro sail blades is simple and the mechanics of blade pitch control have already been developed for use on helicopters. The heliogyro would have a theoretical capability of carrying a spacecraft of over two tons, or about 1500 pounds (600 kilograms) more than the large single-shot solar sail.

The major uncertainty in heliogyro design is the dynamics associated with the 6.25-kilometer blade and its operation in deep space.

SOLAR ELECTRIC PROPULSION

A new rocket engine based on the principle science fiction writers term "ion-drive" is under construction at Caltech's Jet Propulsion Laboratory. The new rocket will produce thrust by using the sun's energy to vaporize fuel, electrically charged by ionization, and accelerate it to tremendous speeds. The rocket's power becomes apparent when we realize it works in a vacuum of space against a nearly negligible gravitational field with a fuel efficiency ten times greater than that of a conventional rocket. The goal is to couple lightweight solar arrays which produce electricity to a cluster of ion engines to propel an instrument package to a rendezvous with Halley's Comet.

When fully extended, the roll-out or fold-out solar arrays would give the ion-drive spacecraft the appearance of a large galactic butterfly, with a potential wing span of about 492 feet (150 meters), the length of 1½ football fields.

Launch of the solar electrically propelled spacecraft for a rendezvous with Halley's Comet would take place by shuttle in 1982. The ion-driven spacecraft would then place its scientific payload alongside the comet (for formation flight or a rendezvous) a little more than 3½ years later, as the comet makes its closest approach to earth since 1910.

Given an equivalent fuel load, an ion rocket could provide a huge increase of a ballistic-powered rocket in the total thrust delivered. The large capability makes ion drive propelled vehicles candidates for heavy payload flights to Saturn and the other outer planets, and for round trips to Mars as well as tours to asteroids and comets. For a mission to Halley's comet, an ion drive spacecraft would use ten to twelve mercury-ion engines of 11.8-inch (30-

Stages of deployment of the ion-driven spacecraft. Sequence is clockwise from lower left.

centimeter) diameter. In mid-1977, the engines had already completed one endurance test lasting longer than 416 days (10,000 hours).

JUPITER ORBITER PROBE

Since man first observed the planets and noted that they followed different orbits from the "fixed stars," there has been strong interest in Jupiter. Galileo stunned the world in 1610 with his discovery of four satellites orbiting Jupiter. He demonstrated that Copernicus was correct: the sun, not the earth, is the center of the solar system. And since then, scientific interest in Jupiter has not slackened.

Scientists often refer to Jupiter and its thirteen or more satellites as a "mini-solar system." They see a set of

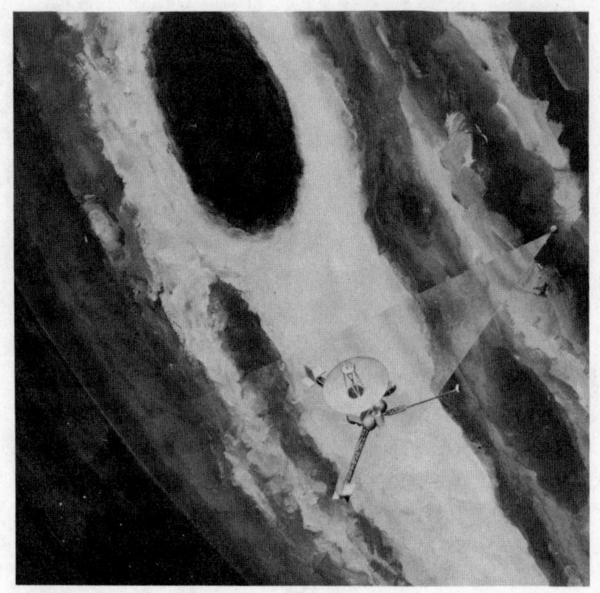

Jupiter Orbiter Probe in the Jovian atmosphere.

similarities between the solar system as a whole and the Jovian system within it. Jupiter appears drastically different from earth and the other terrestrial planets, Mars, Venus, and Mercury. While those planets are mostly composed of rock, Jupiter's main features are hydrogen and helium, in about the same ratio as in the sun itself. Apparently after the sun formed some 4.6 billion years ago, 98 percent of the matter left behind to form the planets went into the creation of Jupiter. Except for the sun, Jupiter is the noisiest source of ratio signals. Its magnetic field is large enough to reach from earth to Venus. Jupiter apparently has no solid surface, but changes gradually from a gaseous hydrogen-helium atmosphere to an interior of liquid metallic hydrogen.

The planet's satellites fall conveniently into three loose groups. The first are the inner five, including Amalthea and the Galilean satellites Io, Europa, Ganymede, and Callisto. The second satellite group appears to be captured asteroids, orbiting Jupiter at great distance. The third group also appears to be captured asteroids, although their orbits are retrograde, in the opposite direction.

The first planetary mission to be carried aboard the Space Shuttle will be the Jupiter Orbiter Probe (JOP), a mission to orbit Jupiter and send an instrumented probe deep into its atmosphere. Designed and constructed by Caltech's Jet Propulsion Laboratory, the mission will allow space scientists to analyze Jupiter for a long period of time at close range. The JOP should provide important evidence on the origin and evolution of the solar system and provide new insights into phenomena directly relating to man's understanding of all the planets, including earth.

According to plans, one spacecraft will be launched in January of 1982 by the shuttle. A flight time of 2.9 years (1,000 days) to Jupiter is planned in order to allow the probe to enter the atmosphere on the sunlit side of the planet. Fifty-six days before JOP arrives at Jupiter, the probe will be released from the main spacecraft.

The instrument-laden probe will offer the first direct sampling of the dense atmosphere of Jupiter. It will descend rapidly through the atmosphere, radioing back scientific findings to the main spacecraft, Orbiter, which will relay them to earth. The probe's scientific payload will weigh approximately 50 pounds (24 kilograms). The probe's mission may be completed as soon as 20 minutes after it first senses Jupiter's atmosphere.

Meanwhile, the Orbiter will burn its rocket engine to settle into a leaping, elliptical orbit around Jupiter and change its orbit so that closest approach will be about the orbit of Ganymede, some 622,800 miles (1,002,400 kilometers or fourteen Jupiter radii) from Jupiter. For the next twenty-five months, the spacecraft will follow a series of elliptical paths that will take it to all regions around Jupiter. This will be accomplished by using Ganymede's gravity to skew the orbit slightly each time the spacecraft approaches Jupiter. The spacecraft will also analyze Jupiter's magnetatail, that portion of the magnetic region directly opposite the sun.

The objectives of the mission are broken into two groups, one for the Orbiter, another for the probe. The probe's scientific objectives include determining the structure (temperature, pressure, and density) of the following:

- the Jovian atmosphere to a pressure depth of at least ten bars, ten times the earth's surface atmospheric pressure
- the chemical composition of the Jovian atmosphere
- the location and structure of the Jovian clouds
- measure vertical energy flux to determine the local radiative energy balance
- characterize the upper atmosphere
- determine the nature and extent of cloud particles

The Orbiter's science objectives are to:

- determine the surface composition of the satellites
- identify the physical state of satellite surfaces and characterize their surface morphology
- measure the satellites' magnetic, gravitational, and thermal properties to obtain their geophysical characteristics
- dissect the satellites' ionospheres, atmospheres and emission of gases
- define topology and dynamics of the outer magnetosphere, magnetosheath, and bowshock
- describe the nature of magnetospheric particle emission
- determine the distribution and stability of trapped radiation
- study magnetosphere-satellite interactions
- conduct a simultaneous observation of the varied Jovian atmospheric properties such as temperature, composition, density, and cloud formations

The JOP spacecraft will feature a new design using the best aspects from two kinds of spacecraft. Part of the spacecraft will be three-axis stabilized while the other will be spin-stabilized. JOP will also incorporate a 16-foot diameter furlable antenna, similar to those on the recent Applications Technology Satellites. Power will be provided by new radioisotope thermoelectric generators, advanced versions of the newly developed nuclear power sources used on previous outer-planet missions.

JOP would be America's third mission to Jupiter. Its predecessors include Pioneer 10 and 11, and Voyager 1 and 2. Pioneer 10 flew past Jupiter in December of 1973, while Pioneer 11 aimed one year later. Voyager 2 was launched to Jupiter in 1977, while Voyager 1 was launched 12 days later. They will arrive at Jupiter in March and July of 1979 on their way to Saturn.

Pioneer II passing Saturn in September 1979.

VOYAGER JUPITER-SATURN

Mission plans call for the first-launched Voyager to fly a slower trajectory allowing the second to overtake it and reach Jupiter about four months earlier. Jupiter's gravity will sling-shot them toward the ringed planet Saturn. By the time the spacecraft reach Saturn, the leader will be about nine months ahead.

Jupiter and Saturn are drastically different from the terrestrial planets, since both appear to be composed of hydrogen and helium. Jupiter is larger than all other objects in the solar system combined except for the sun. In a small telescope, Saturn's unique rings appear dazzling. Jupiter has thirteen or fourteen satellites (the recently discovered fourteenth has not yet been confirmed). Saturn has ten satellites, including Titan, which has an atmosphere more dense than that of Mars. Jupiter orbits the sun more than five times farther away than earth. One Jovian year equals 11.86 earth years and Jupiter's day is less than ten hours long.

On the other hand, Saturn orbits the sun almost ten times as far away as earth. It completes one orbit every 29.458 earth years. A day on Saturn is 10 hours, 26 minutes long. The widest visible ring has a radius of 85,000 miles (137,100 kilometers).

Photography of Jupiter will begin eighty days before Voyager reaches the planet, along with spectral scans of the hydrogen cloud surrounding Jupiter and the orbits of the four big Galilean satellites. Images of the brightly banded planet will already exceed the resolution of earth-based photographs. For approximately two months photography will continue with the spacecraft's narrow-angle camera, which has a 1,500 mm focal-length telescope. Eight days from Jupiter, Voyager will begin coverage of the entire planet with its wide-angle camera (200 mm focal length), while the narrow-angle instrument provides high-resolution photography of selected features of Jupiter's clouds. At the

Diagrammatic illustration of Voyager flight to Jupiter and Saturn.

same time, the infrared and ultraviolet spectrometers and the photopolarimeter will be obtaining data on atmospheric compositions, temperature variation in the atmosphere, and the possibility of solid particles in the clouds.

Shortly before its closest approach to Jupiter, Voyager will fly within 273,000 miles (440,000 kilometers) of Amalthea, giving scientists their first close look at the innermost of Jupiter's satellites. Closest approach to Jupiter will be five Jupiter radii (R_J) from the center of the planet (some 222,000 miles or 357,000 kilometers). Jupiter will occlude the sun and earth, allowing scientists to make precise measurements of the structure and composition of its atmosphere.

After passing Jupiter, the first Voyager will examine all four of the big Galilean satellites: Io from 15,000 miles (25,000 kilometers), Europa from 466,000 miles (750,000 kilometers), and Ganymede and Callisto from 81,000 miles (130,000 kilometers). Observations of Jupiter will continue for approximately a month after closest approach.

The second Voyager will begin its observatory phase about two weeks later, again eighty days before closest approach. It will observe four satellites during the inbound leg: Callisto from 150,000 miles (240,000 kilometers), Ganymede from 31,000 miles (50,000 kilometers), Europa from 118,000 miles (190,000 kilometers), and Amalthea from 342,000 miles (550,000 kilometers). The spacecraft following a more distant path than its predecessor will pass 10 R_J from the center of the planet (443,000 miles or 714,000 kilometers).

A few months later, the first Saturn encounter will begin and continue for several months. On the inbound leg Voyager will pass within 4,000 miles (7,000 kilometers) of the major satellite Titan. It will also scan the moons Rhea, Tethys, and Enceladus before passing Saturn and its rings. The closest approach to Saturn will occur at 130,000 miles. Titan, Saturn, and the rings will occlude the sun and the earth as seen by instruments on the spacecraft. The second Saturn encounter will begin in the summer of 1981. Closest approach to Saturn will occur a few months later. The spacecraft will observe the satellites Titan and Tethys on its inbound leg, and Euceladus, Dione, and Rhea after closest approach.

The Voyager spacecraft will weigh 1,753 pounds (795 kilograms). The scientific instruments weigh a total of 220 pounds (100 kilograms). The new spacecraft differ from Mariner planetary craft, due primarily to the environment into which they will venture and the great distance across which they must communicate with earth. Since the outer planets receive only a small fraction of the sunlight that strikes earth and Mars, the Voyagers cannot depend on solar energy but must use nuclear power, radioisotope thermoelectric generators. Another obvious difference is the large antenna: the antenna on Voyager is 12 feet (3.7 meters) in diameter. Each Voyager will use ten instruments and the spacecraft radio to study the planets, their satellites, the rings of Saturn, the magnetospheres surrounding the planets, and interplanetary space.

In addition to wide-angle and narrow-angle television cameras, the Voyagers will carry cosmic ray detectors, infrared spectrometers and radiometers, low-energy charged-particle detectors, magnetometers, photopolarimeters, planetary radio-astronomy instruments, plasma and plasmawave experiments and ultraviolet spectrometers. The television cameras are expected to provide scientists with pictures of Jupiter and Saturn that are clearer than have ever been seen, and the first high-resolution close-up images of the Galilean satellites of Jupiter, the major satellites of Saturn, and Saturn's rings.

Other instruments will probe the atmospheres of the planets and satellites, their magnetospheres and the interactions between these regions, and the solar wind and radio bursts from Jupiter. Other objectives include occluding earth and sun by the planets, Saturn's rings and Titan, all-sky surveys of interplanetary space, and location and definitions of the heliosphere or boundary of the solar wind.

Trajectories were carefully selected to provide not only good scientific information about the planets, but also about their satellites. If Io, Europa, Ganymede, Callisto, and Titan orbited the sun instead of planets, they would qualify as objects for close analysis in their own right. They range from larger than the planet Mercury down to the size of earth's moon. Titan is the only satellite in the solar system known to have an appreciable atmosphere.

Space scientists have the option to send the second spacecraft on to the planet Uranus with the encounter taking place in the spring of 1986. The Uranus option will be exercised only if primary Saturn science objectives have been met by the first spacecraft and the operating health of the second warrants such an undertaking. Because of the alignment of the outer planets, there is also a possibility that the second spacecraft could be targeted to Neptune.

At planet encounter, high-rate data will be received

through the DSNs 210-foot (64-meter) antenna subnet. Maximum data rate at Jupiter is 115,000 bits per second. At Saturn, it is 44,000. The cost of Project Voyager, exclusive of the shuttle launch vehicle and operations, is $338 million.

THE PUBLIC SERVICE PLATFORM

As long as thirty-five years ago, Arthur C. Clarke, co-author of the book *2001: A Space Odyssey,* suggested that a satellite be positioned some 23,000 miles above the equator where it would be stationary. Designed and constructed as communication relays, three of these geostationary satellites would be able to "see" virtually the entire globe. Thirty-five years ago, experts rejected the idea as "impossible." Today, NASA's chief administrator Kenneth Fletcher practically takes these geostationary communication satellites for granted. They are positioned and function almost routinely, exactly as Clarke envisioned.

Fletcher notes that less than a decade ago experts doubted whether space scientists could really place a man on the moon and were calling the proposed exploration of Mars an improbable dream. At the time, the notion of providing easy, inexpensive access to space from a cargo-carrying, reusable shuttle was beyond belief.

Today space scientists feel that the average person will be the main beneficiary of the new space age, an age in which earthlings will be able to conduct industry, science, and even a wide range of public services from vantage points in orbit.

After carefully narrowing down NASA's list of some two hundred possible public service functions to thirteen first-priority functions for a satellite relay, space engineers at Grumman Aerospace have envisioned an integrated triple-antenna satellite, or Public Service Platform (PSP). The idea revolves around the positioning of a huge multi-beam antenna with very high microwave-radiated power in geostationary orbit.

MULTIPLE-FUNCTION USER TERMINALS

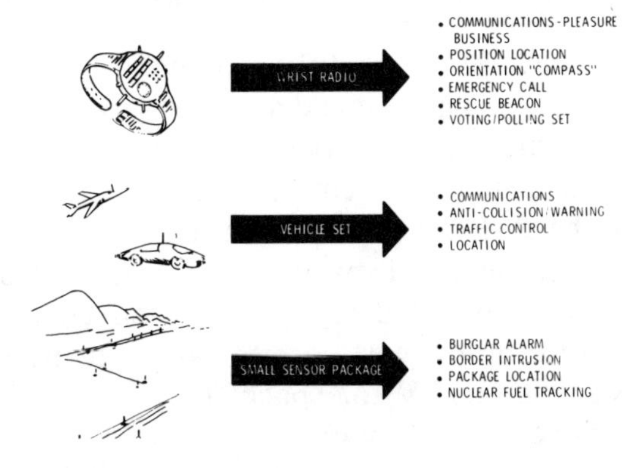

GUIDING PRINCIPLES FOR SPACE APPLICATIONS

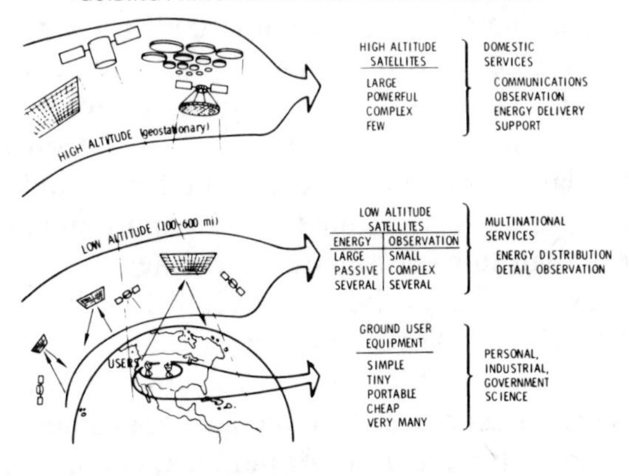

MULTIPLE-FUNCTION SATELLITE TYPES

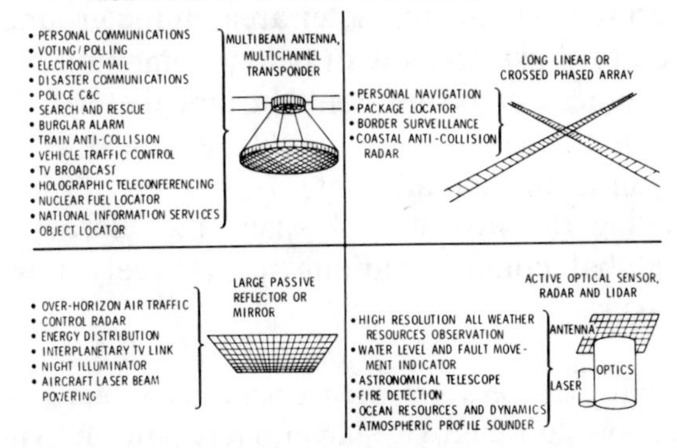

Public Service Platform.

A structure as large as a PSP antenna would be lifted in sections into low-earth-orbit and assembled. Early flights of the Space Shuttle would deliver hardware into space for initial structural model assembly and performance tests. Then, once the PSP design had been further evaluated and perfected, a large manned space station, a Space Construction Base, would be used to assemble and check out an operational PSP in low-earth-orbit. Later, an Inter-orbital Transfer Vehicle (IOTV) or space tug, would propel the PSP up to its geostationary orbit.

To Grumman engineers, the best possible proving ground would be the Space Construction Base (SCB), with its own living quarters, the "foundation" for a permanent space colony. Along with carrying the Public Service Platform material to low-earth-orbit, the shuttle would support the construction and safety of the Space Construction Base. Calculations demonstrate that the PSP could perform the thirteen public service functions for approximately $750 million less than thirteen separate satellites.

The design calls for the first antenna in a three-antenna PSP to be almost 200 feet in diameter and comprised of some 150,000 elements. It would function with some 44 kw of dc power and generate 256 fixed and 16 scanning beams. If this 30-ton PSP were used for personal communications, one of the thirteen functions studied sets of spot beams each covering a 60-mile diameter area on the ground would be devoted to those areas with heavy communication traffic. For instance, six beams could be aimed at the northeast corridor, another six beams could be aimed at the Los Angeles-San Francisco area, etc.

Recalling that the use of space has already revolutionized global communications in relatively few years, Fletcher asks:

> If public service users need small, handy earth stations, how do we provide larger, more powerful satellites with enough channel capacity to make the cost

Diagrammatic illustration showing deployment of communication satellite.

per channel affordable? Until now, that has been a permanent concern. But things are changing. Use of the shuttle will reduce the cost of placing a satellite in orbit from more than half to less than a quarter of the total cost of design, construction, and launch. Indeed, the shuttle will make it possible for communications satellites to have multiple frequencies operating at high power, with the satellite positioned very accurately and remaining in service on station for a long time. The implications for public service are spectacular.

The communications satellites may provide services and generate products that are superfluous and no more than sophisticated toys. Some, such as the "Dick Tracy wrist radio," do have practical value as well. Naturally, many factors such as economic viability and acceptance by the public, industry, and political sectors would have to be considered before such a universal communications scheme could be implemented. But, the readiness of the technology and an admittedly intuitive feel for the need for such a service led to its espousal by scientists and engineers. Their analyses show no reason why a $10 transmitter-receiver (transceiver) could not be worn on the wrists of some three million Americans (probably controlled and administered by the telephone companies). Thus, people could converse from coast to coast any time they desired for a given fee. Their calls would be routed and billed automatically.

The rechargeable transceiver would weigh less than two ounces and function on a fraction of a watt of power for twenty continuous hours. Upon receiving a person's voice transmitting his coded telephone address as well as that of the person he wants to talk to, the PSP would almost instantaneously trigger a buzzer in the target transceiver, signaling the two-way radio-telephone conversation to begin.

Current space communications systems tend to use relatively small, inexpensive satellites which can communicate with only a few large, fixed, and expensive earth terminals. The Public Service Platform would change all this. Placing the high-power, large-aperture antennae along with the big switching centers in space instead of on earth would provide relatively inexpensive, high-quality communications free of the fading and interferences which often frustrate earthbound radio systems. Engineers feel that the required technology, already in existence, will enable them to experiment with the PSP in space on a small scale during 1985.

Grumman's Ross Fleisig, Space Systems Engineering Project Leader, and Joel Bernstein, Engineering Group Specialist, offer several examples of how the wrist radio could help man. For instance, if a hunter is accidentally hurt in the wilderness, he or his companions can radio for help. Another area which would benefit from the PSP is law enforcement. Today, police radio communications lack the required degree of security and reliability. Over wrist radios, police communications would be free of jamming and eavesdropping. Still another benefit revolves around election voting or polling. By means of wrist transceivers, it would be possible to poll or record the votes of 100 million people in one hour, allowing many more Americans to participate in the shaping and working of their democracy. If a Congressman wanted to poll constituents on an issue before Congress, he could simply "beep" them all simultaneously and ask for an immediate response by mail. Similarly, a businessman or advertiser could get a fast sampling of a product's usefulness or appeal to potential customers.

Fleisig and Bernstein say that based simply on the projected growth of conventional communications, telephone, television relay, etc., PSP shows clear potential for becoming a multibillion-dollar space industry by the late 1980s. The telecommunications industry is expected to become heavily involved, just as it has in communications satellites. Already, satellites enable us to communicate overseas at half the cost of conventional undersea cable and radiotelephone networks.

Grumman engineers also insist that the universal wrist radio could benefit disaster control. For example, floods, earthquakes, fires, etc., will often destroy normal communications as well as overload emergency radio channels. Further complicating such emergencies, much additional equipment rushed into service is often incompatible with each other and with normal communications media. The Public Service Platform would be built with fixed radio beams, one for each of 260 urban command-and-control centers across America, each carrying 16-voice channels. Also, twelve steerable beams could be aimed at many disaster areas simultaneously. Rescue and relief workers carrying portable transceivers would have two-way communications with each other and with their urban control center. People, equipment, and supplies could be dispersed much faster and more effectively to save lives. Engineers say the same concept also could be used for civil defense or other national alerting systems.

Another benefit of the Public Service Platform will be electronic mail. A similar antenna system, erected by a manned Space Construction Base, could take care of more than 30 percent of the present load of first-class mail, saving taxpayers nearly $1 billion a year. With the mail growth rate expected to continue its sharp climb, the post office probably will have to invest a large amount of additional capital just to keep up. This NASA-developed supplementary mail system could be one way out. The plan calls for page readers and facsimile printers to be installed at each of some 100,000 U.S. post offices to read designated mail and transmit the contents to the PSP antenna for relay to destination post offices, where they would be reconstituted. Postmen would continue to deliver postage between home or business and the post office, but transportation of this mail from one post office to another would be eliminated, along with a tremendous amount of handling and sorting.

Fully automated reading, printing, and sealing machines would preserve the privacy of the mail. An added precaution would be to use a secure digital code system, involving the scrambling and unscrambling of messages with procedures similar to those used today by military and government agencies.

The possible offshoot of this electronic mail concept could provide a national information service. Various individuals, privately or in business, often need the same kind of information all year around. Examples include Veteran's Administration data, Library of Congress documentation, medical and other libraries, various data banks, and archives. If such sources of information were made accessible in a centrally located computer information bank, a user could obtain information through the PSP in a matter of minutes.

Again Fleisig and Bernstein outline possible examples:

> One can envision the value of satellite service to an easterner who becomes ill while visiting the west coast, 3,000 miles away from his doctors and medical records on the east coast. Obtaining such vital data today can be extremely time-consuming and difficult. But if the needed information were stored in a central information file, the Public Service Platform could re-

> trieve it rapidly enough to save lives. The central file also would make it simple to rapidly update the information by inserting new data. Again, privacy of the information would be preserved by coding. Also, were the PSP disabled or destroyed in orbit, the data would still be preserved in the ground-based computer.

Still another Public Service Platform function would be detection and control, which could help prevent diversion or hijacking of nuclear fuel, thereby minimizing the danger of nuclear terrorism, blackmail, and proliferation of nuclear weapons. At the time nuclear material is produced, a radio transmitter will be built into the fuel. Such a gadget would emit a coded signal continuously and the authorities could maintain constant scrutiny of the movement of the nuclear material. Monitoring devices placed in reservoirs and automatically interrogated by PSP radio signals from time to time could keep scientists informed on the availability of water resources.

Because a platform in space can measure thousands of points on earth in a few moments, it will be possible to obtain earthquake patterns nearly simultaneously. With finely detailed individual observations, the PSP could be used for earthquake detection and prediction. As many as a million seismic sensors, installed along faults and other suspect areas throughout the world could be sampled hourly. This would provide the long-sought means of monitoring earthquake onset over the entire world at one time. Seismologists say that the number of suspect areas makes it uncertain as to which might become active next, so it is not enough simply to monitor the few earthquake regions that have been the most active historically. The PSP would spot the onset of any earth motion almost immediately and signal an alert.

The ability of the PSP's sensors to monitor such large numbers of sensors at the same time suggests an effective regional or national burglar alarm/intrusion detection system. Conventional pressure-actuated sensors placed in lawns and sidewalks would be triggered when doors or windows are opened during inappropriate periods. These sensors would be linked to tiny transmitters in critical governmental installations (such as prisons, mints, and headquarters complexes), industrial buildings, and even private homes. It would be entirely feasible to process up to 6,000,000 intrusions per second within each of 500 major urban areas, while also monitoring some 2.5 billion intrusions per second elsewhere throughout the country. Within a fraction of a second of the unwanted entry, the PSP would automatically alert a police receiver nearest that point.

Because television signals travel in a straight line, there are still many mountainous, rural, and remote regions of the U.S. that have either poor television reception or none at all. These "deprived" areas could be eliminated by television signals beamed from a single Public Service Platform in orbit.

Grumman engineers also recommend serious consideration of one further use for a PSP—holographic teleconferencing. By using established techniques of laser holography, the PSP could provide TV coverage of business meetings with three-dimensional picture quality. Such a PSP service might save billions of dollars a year in airline travel, room and car rentals, and the cost of the time spent by business people in transit.

Space engineers envision hundreds of widely separated conference rooms equipped with conventional communications aids—television camera and projector, laser illuminator, and stereo sound system—to produce three-dimensional, life-size images of the conferees. Brought virtually "face-to-face" by the PSP system, they would be able to perform almost all business functions except the handshake. The 3-D system would be particularly useful in illustrating working models of equipment, reviewing data, and in gauging reactions on all sides to the business propositions being presented. For less sophisticated requirements, a businessman might elect to use cheaper, two-dimensional black-and-white or color TV.

Final proof of the feasibility of the public service concepts would come from actual hardware demonstrations in orbit.

THE SPACE FACTORY

The vision of constructing factories in space has appealed to space scientists and officials as much as the concept of a Public Service Platform. Experimental data gathered in Biosatellite II in 1967, Skylab in 1973, Apollo-Soyuz (the joint US-USSR venture) in 1975, and other orbital flights are being used by NASA and aerospace contractors to select products which might be manufactured more efficiently in the near-zero gravity hundreds of miles above the earth.

In studies partially funded by NASA's Marshall Space Flight Center, space technologists have defined two representative manufacturing operations for space. After weighing more than two dozen promising operations against such criteria as potential benefits, range of applications, costs, technological difficulty, and availability of previous experimental results, the list was narrowed down to two promising products: Urokinase (an important pharmaceutical) and permanent magnets.

Urokinase (yoor′-oh-ky′-nays), an enzyme found in minute quantities in human urine, has the extremely beneficial property of dissolving blood clots. Until very recently, it could be produced only from large quantities of human urine, which both limited the amount that could be produced and magnified its cost. Recently, the enzyme has been selling at about $1,500 for each treatment of pulmonary emboli—blood clots which may enter the lung, usually after surgery.

The mortality rate from embolic surgery is about 68 percent, and an estimated 15,000 persons die a year from urokinase-treatable diseases in the U.S. alone. Another 100,000 non-fatal cases per year are potential candidates for urokinase therapy in America. At least one pharmaceutical manufacturer has developed methods to produce urokinase more efficiently from living cells grown in tissue culture and plans to begin marketing the resulting enzyme product abroad this year. In 1977, the firm asked the U.S. Food and Drug Administration to approve the use of the enzyme for treatment of pulmonary embolism in the U.S. According to Gary Geschwind, laboratory head of Materials Research at Grumman Aerospace, packaged urokinase would cost about one-tenth as much to produce in space as on earth—with the "spacekinase" eventually selling for $75–$100 per dose (including the cost of getting the material to and from space).

Experiments flown on Biosatellite II indicated that microbiological propogation rates in space may be greater than on earth and that yields also may be greater. Grumman's own tissue-culture analyses as part of its space station studies hinted at increased yields of urokinase in space and thus lower costs. These results have taken on ever greater significance in light of the results of the U.S.-Russian Apollo-Soyuz Test Program experiments to separate human embryonic kidney cells by an electrical process. After the separated, frozen material was returned to earth, it was fractioned into 30 segments and then tissue-cultured. One culture had produced a tenfold increase in yield of urokinase over comparable control materials produced on earth.

Although scientists caution that this represents only a single result, it is considered very encouraging because of the implication that electrical (electrophoretic) separations of urokinase as well as other biologicals could drastically reduce the cost of insulin, antihemophilic factor, growth hormone, and other drugs.

Analysis of substances which cannot be mixed in low gravity also indicate that tissue culturing can be improved by eliminating contact inhibition, that is by removing the conditions which retard the growth of cells due to their contact with one another.

On earth, tissue-cultured cells must be spun rapidly in a flask to deter the natural sedimentation and to maintain them in suspension. Because they are very fragile, many are lost, or their yield reduced, due to the turbulence in the flask. With the absence of sedimentation in the low gravity of space, there would be no need for spinning. Thus, it may be possible to select cells which would grow in suspension, unattached, from among cell populations that grow only as a monolayer on earth.

Separation of urokinase-producing kidney cells from other non-producing kidney cells can also be done with much greater precision in space because of the almost total absence of thermal convection, the heat transfer which inhibits growth that takes place because of natural motion of cells in non-solids. An absence of thermal convection would increase specific separation of cells.

The first step in this process is free-flow, or continuous (as opposed to static) electrophoretic separation. The second step is tissue culturing, with which scientists expect to produce much higher yields of urokinase in space than on earth as a result of the absence of gravity-induced sedimentation as well as growth-inhibiting contact among cells. Studies indicate that producing in space about 4½ pounds (two kilograms) of urokinase a year to satisfy a U.S. market is a feasible operation. A pilot manufacturing plant could begin in 1985, with commercial manufacturing operations getting underway two years later with FDA approval.

The selection of permanent magnets for manufacture in the space factory resulted from the experiments carried out in the 1975 Apollo-Soyuz space laboratory. The particular experiment was one designed by materials scientist David Larson of Grumman's research department. Larson's Apollo-Soyuz experiment processed magnets of a manganese bismuth alloy in an ingenious miniature space laboratory. The experiment produced materials with magnetic properties well in excess of that ever produced on the ground. In some instances, Larson discovered a 100 percent increase in the coercive strength (attraction-repulsion force) of magnets. On the basis of these findings and continued research, Larson and his colleagues are now highly enthusiastic about the prospect for manufacturing permanent magnets in space.

With the availability of permanent magnet motors with up to 50 percent greater efficiency than today's units, the demand for them is expected to steadily increase. A major feature of the space-manufactured units would be much greater resistance to demagnetization and loss of strength when overloaded. This would be of great advantage in today's automobile which uses as many as twenty different motors (starter, distributor, heater blower, etc.) and other electrical devices employing magnets. Space "permagnets" also show promise of significantly reducing the weight and operating costs of battery-powered vehicles as well as increasing their operating range.

The applications for space-manufactured magnets are far-reaching. Space magnets could reduce the weight, unreliability, and possibly the cost of such common household appliances as vacuum cleaners, drills, and sewing machines. Advantages could extend to the medical field in such uses as lightweight orthopedic braces, heart pacemakers, devices to control urinary functions, and many other areas. Interestingly, recent studies by European physicians indicate that scar tissue is greatly reduced or eliminated when wounds heal in the presence of strong magnetic fields.

As magnets age, their properties may change. This is a problem in navigational instruments, which require a high degree of accuracy. This requires that they be periodically recalibrated to retain accuracy. Space manufacture, however, would provide inertial navigation magnets with far more enduring properties.

Considering their potential benefit to the Air Force alone, technologists believe the use of these more stable magnets could save about $50 per flight hour, or at least $42 million annually. Other areas that could be similarly benefited include computer line printers and other peripheral printing devices; loudspeakers for sophisticated music systems; and even typewriters, the ball rotators could be made simpler, lighter, and less costly.

Capable of producing 20,000 pounds of permanent magnets a year at a wholesale price of $1,500 per kilogram, the initial orbiting factory would consist of a habitation module and a general-purpose laboratory module with manufacturing equipment. The machinery would be mounted in pressurized cannisters outside the lab module. Inside, rods of pressed cobalt/rare earth alloy powder would be melted in a quartz lamp furnace, where they would be suspended in a containerless high-purity inert gas. This would keep the material from picking up oxygen, nitrogen, carbon and other impurities from the container walls as in the case of conventional processing on earth. Such impurities cause the magnets to change properties with use. The space manufacturing process would also permit the nearly parallel and continuous alignment of thousands of feet of microscopic rods, enhancing their magnetic capabilities by perhaps as much as 100 percent in comparison with earth methods. Without gravity, the "gloating" molten material would need no container and the resulting material would be machinable, uniformly dense, and storable in long continuous lengths of spools.

Using a crystal growth process, individual high-quality solar cells or a solar array fabricated from silicon semiconductor ribbon grown by a continuous process also could be manufactured in orbit for both space and terrestrial applications. For example, these more efficient materials could be used in the power units that would beam the sun's energy to electric power receiving stations on earth or to a space factory or Public Service Platform.

THE SPACE CONSTRUCTION BASE

The enormous structures involved in constructing such large-scale orbital facilities as a space "mini-factory," Satellite Power System, and Public Service Platform dwarf the 53 × 15-foot cargo bay of the Space Shuttle Orbiter. Nevertheless, the Orbiter figures importantly in NASA's early space construction concepts.

In the first structure-fabrication experiments, this spaceplane will be used as a working platform. Subsequently, it may serve as part of the transportation system operated in support of a full-scale Space Construction Base (SCB), the initial site of man's permanent occupancy.

Grumman Aerospace scientists Joe Marino and Al Nathan recently completed a nine-month study for NASA's Lyndon B. Johnson Space Center of an "Orbital Construction Demonstration Article" (OCDA) which could be the forerunner of the Space Construction Base. They analyzed the design of a 236 × 105-foot platform in low-earth-orbit, stabilized around three axes by flywheels and equipped to make orbit-keeping adjustments. Orbit-keeping is required to make certain that the OCDA remains

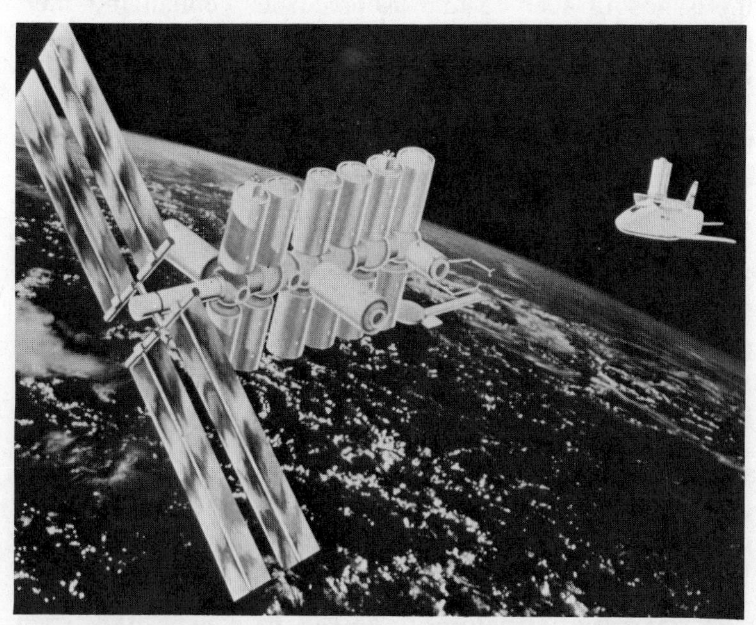

Orbiter approaches temporary space base carrying another section to be plugged into the core.

at the required 210-mile altitude with an error of no more than 10 miles in a six-month period. Engineers estimate a prototype OCDA could be assembled in three Orbiter flights over a three-month period. The first Orbiter flight would establish a platform that could operate autonomously and be stable enough for Orbiter docking.

The studies by Marino and Nathan were aimed at defining the requirements future ultra-large space structures will impose on near-term construction technology. Some forty future orbital missions were reviewed, thirty of which were eliminated because they didn't require space construction. The remaining ten missions were halved by eliminating systems with identical requirements. The five representative missions selected were: Satellite Power System; Microwave Power Transmission System; Radiometer Reflector; Solar Thermal Power Satellite; Solar Mirror.

Most initial construction operations will be automated or under remote control. After the Orbiter's big cargo bay doors are swung open in flight, one of the four members of the construction crew takes a post on the spaceplane's flight deck at the control panel of the Orbiter's remote manipulator system. This long, hinged arm with a pincer-like device at its free end lifts the core module/mast out of the cargo bay and rests its docking ring on the Orbiter's docking ring.

The three other members of the crew, dressed in Extravehicular Activity (EVA) spacesuits, lock the rings to each other, aided by the remote manipulator systems operator in the shirtsleeve environment of the flight deck. The mast now provides a passageway in and out of the docked Orbiter, the only living quarters provided with the OCDA. Next, the boom/solar array drive unit is mounted on the end of the square mast. This initial solar array is of modest size, compared to the final one. It provides 19.2 kw of "housekeeping" power for sustaining platform functions between Orbiter visits, while also powering a moderate experiment load. The array tracks the sun as the platform orbits the earth once every 1.5 hours.

Studies indicate that during the first seven-day mission, it would take four men working one shift—no more than 5.75 EVA hours per day—about 120 manhours to assemble the core/mast and install the reaction control system and antennae. Following this, the crew would return to earth.

Shortly thereafter, a second one-week shuttle flight would deliver a seven-man crew to construct the rotating boom, complete the solar array, and build a major portion of the platform. This would be done in two shifts, or a total of about 240 manhours.

It will take two seven-man crews, one working two shifts on the second Orbiter flight and one working two shifts on a third, 547 manhours to assemble the OCDA. When completed, the facility would weigh about 82,000 lbs.

Left vacant in space between Orbiter flights, the Orbital Construction Demonstration Article, or platform, would ultimately be succeeded by a permanent Space Construction Base (SCB) in a 300-mile-high orbit which would support four to six men at a time, with crew changes every four months. Each of the SCB's two larger modules, measuring 13 × 43 feet and weighing a maximum of 53,000 pounds, requires one Orbiter flight. Joined together, they will house five people, who will perform such tasks as monitoring prototype space-manufacturing processes, performing engineering tests on material samples, and fabricating beams for simple structures. A smaller logistics module will carry enough supplies for three months. When it runs out, the spent module will be returned to earth by the Orbiter, and a fresh module attached. The second large module is a general-purpose laboratory which houses a prototype space manufacturing plant for biological, crystal growth, and solidification processes. It also provides facilities for material, joint, and mechanical tests. Powering both modules is a two-winged solar array that provides full power in sunlight, with storage batteries providing power during dark periods.

A two-person external airlock permits extravehicular

activity. A transparent dome provides a shirtsleeve environment for the two people who operate a pair of manipulators. These could handle and assemble triangular 3.3-foot (one meter) beams provided by a fabrication module (spacelab) attached to the outside of the lab.

To avoid the costly transportation of bulky, awkward structures into orbit, space engineers have been studying ways of automatically fabricating these beams in space, which would collapse under their own weight on earth. Recently, engineers developed a prototype one-meter trusswork beam, using six commercial rolling mills to form the aluminum cap and shear members. They are currently constructing a beam made of an advanced composite material which shows promise of resisting thermal deflections. On a Space Construction Base, using these basic structural elements, the material would feed through a system of rollers and be automatically spotwelded to emerge from the "spacelab" module as finished one-meter beams. In the case of a solar power demonstration platform, these beams could reach lengths of 1,320 feet.

IV

Celestial Cities: Colonization of the Solar System

"The next step is out there. Out there stars shine, pieces of light . . . a pattern of so much brilliance that I am honored even here."

Lt. Col. Alfred M. Worden

Although thoughts about migration into space are as old as science fiction, the technical basis for serious calculation did not exist until the early 1970s. Just a few years ago, such questions as whether or not scientists could place a man on the moon or construct permanent communities off the earth would have been dismissed as pure fantasy, or at best the wishful projections of scientists several centuries ahead of their times. Today, however, these questions are asked seriously—not only because of curiosity, but also because celestial cities offer potential benefit and hope to an increasingly enclosed and circumscribed humanity.

Scientists feel that man can indeed colonize space today, and accomplish this without destroying or polluting anything. From their calculations, it appears that if work is begun soon, nearly all man's industrial activity could be moved away from the earth's fragile biosphere within seventy-five years. The technical imperatives of this migration of people and industry into the universe are likely to encourage self-sufficiency, small-scale governmental units, cultural diversity, as well as a high degree of independence. Some scientists suggest that the ultimate size limit for the human race on the newly available frontier is at least 20,000 times its present value.

The main focus of a space colony is an environment where some 10,000 people would work, raise families, and live out their normal human lives. Such a habitat would appear wheel-like in structure. This celestial city would orbit the earth in the same orbit as the moon in a stable position equidistant fron both earth and moon. Space scientists who developed the concept refer to this as the Lagrangian libration point, L_5.

The colony would consist of a tube 427 feet in diameter bent into a wheel over one mile in diameter. Residents would live in the ring-shaped tube which would be connected by six large access routes, or spokes, to a central hub where incoming spacecraft dock. These spokes would be 48 feet in diameter and provide entry and exit to the living and agricultural areas in the outer tubular region.

These spokes contain the power cables and heat exchanges which connect the interior of the colony to the external power supplies and the radiator. They also function as elevator shafts through which several thousand commuters travel every day to and from their work in the fabrication sphere or outside the settlement.

The habitat's interior would be illuminated with natural sunshine. The sun's rays in space would be deflected by a large stationary mirror suspended directly over the hub. Such a mirror would be inclined at 45° to the axis of rotation. It would direct the sunlight onto another set of mirrors which in turn would reflect it into the interior of the habitat's tube through a set of lowered mirrors designed to admit light to the colony while acting as a baffle to stop cosmic radiation. A cycle of day and night is necessary in

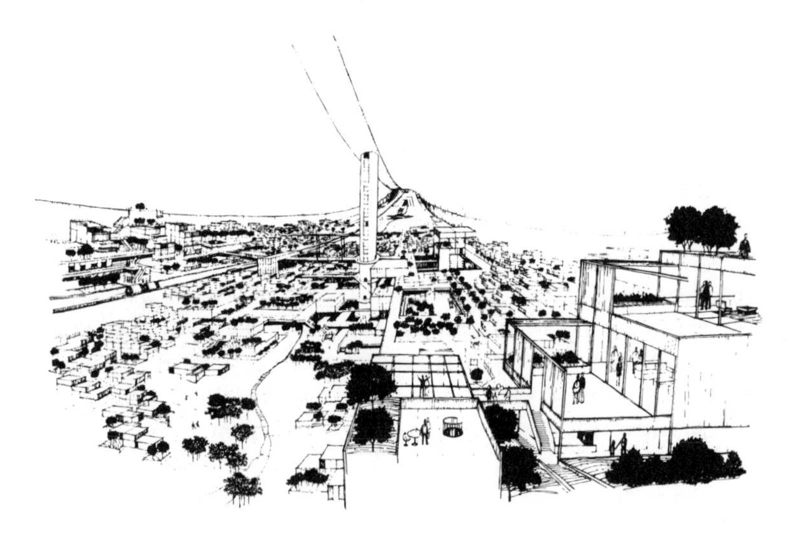

residential as well as some agricultural areas, while other arricultural regions need continuous solar illumination. This is taken care of by reflecting the light away from certain windows to obtain darkness and by focusing the light from several huge mirrors onto other windows. With the help of abundant natural sunshine and controlled agriculture, the colonists (10,000) would be able to raise enough food for themselves on only 160 acres of soil.

Engineering scientists say that abundant solar energy and large amounts of matter from the moon are keys to successfully establishing a community in space. Not only does the sunshine foster agriculture of unusual productivity, but also it provides energy for industries needed by the colony. Using solar energy to generate electricity and to power solar furnaces, the colonists would refine aluminum, titanium, and silicon from lunar ores shipped inexpensively into space. With these materials, they would be able to construct satellite solar power stations and new colonies. The lower stations would be placed in orbit around the earth, to which they would deliver valuable electrical energy. The economic value of these power stations would go far in justifying the existence of the colony and the creation of even more colonies.

Such a system has been conceived to meet a set of specific design goals established to guide the choice of the principal elements of a practical colony in space. The main goal, as scientists see it, is to design a permanent community in space which is sufficiently productive to maintain itself, and to use in a creative way the environment of space to an extent which will permit growth, replication, and the eventual creation of even larger communities. This initial community is designed to be the very first step in an expanding colonization of space.

In order to achieve this main objective, NASA space scientists at the Ames Research Center in Mountain View, California, have established the following goals which must be met using existing technology at a minimum of cost:

1. A habitat to meet all the physiological requirements of a permanent population and to create a viable social community;
2. Obtain an adequate supply of raw materials and produce the means by which they can be processed;
3. Provide an adequate transport system to carry people, raw materials, and items of trade;
4. Develop commercial activity sufficient to attract capital and to produce goods and services for trade with earth.

GERARD O'NEILL'S EARLY CONCEPTS

Active interest in space colonization as a definite, practical possibility began in 1969 when Professor Gerard O'Neill and his graduate students at Princeton University began seriously assessing the feasibility of space colonization. Their assignment was to create a model which would demonstrate how a space colony could easily be created. Work-

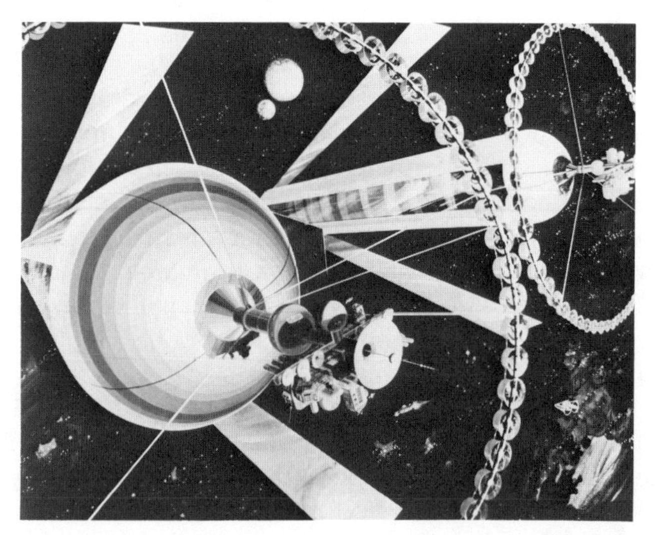

An artist's conception of one of O'Neill's space colonies.

Interior view of the O'Neill concept.

ing over a period of a year, they designed a rotating habitat in satellite orbit around the earth at the distance of the moon, using solar energy to sustain a closed ecological system. They conceived an environment constructed of processed lunar ore delivered by an electromagnetic accelerator. The habitat was seen as a 1-km long cylinder with hemispherical end-caps. It was to have an earth-like internal environment on the inner surface and be supplied with sunlight reflected from mirrors.

A short time later, the Princeton group suggested that the L_5 colony could construct solar power stations from lunar material. They suggested that this would clearly improve the economics of both the satellite solar power stations and the colony itself.

The concept of satellite solar power stations has received increasing attention since its introduction in 1968 by Peter Glaser. But the real credit for the detailed program of space colonization goes to physicist O'Neill. He envisions space as a medium rich in matter and energy which could furnish almost all the raw materials (the moon alone has a tremendous abundance of aluminum and titanium) for a new era in history. His space colony concept proceeds from two premises: that man should create new worlds instead of colonizing old ones, and that we should build them with materials mined from space itself.

Such visions of self-perpetuating worlds in space are not coming from the imaginations of science-fiction writers. O'Neill's original concepts were further developed during a conference in early May 1975, which took place at Princeton University. Entitled "Space Manufacturing Facilities," the conference presented a rationale for possible design choices. Scientists focusing upon O'Neill's test case attempted to detail how the various parts of the habitat would interrelate and support each other. Since the conference, NASA/Ames-Stanford study groups have worked out how the various colony segments would interrelate and support each other. They have established the habitat properties that a successful design must satisfy, what human needs are to be met if people are to occupy such environments, and the specific characteristics of the various alternative components of the design.

HISTORY OF SPACE COLONIZATION

The history of space colonization concepts extend beyond Princeton physicist Gerard O'Neill and back into the legends and myths of ancient times. The first real discussion of an actual space environment appeared in 1870 when American novelist Edward Everett Hale's "Brick Moon" described how a city in space developed by accident when a brick sphere, intended for orbit to guide maritime navigators, was accidentally rolled onto the catapult with workers and engineers still inside.

According to space science historian Robert Solkeld, precursors of the notion of small self-contained worlds in space appeared in novels by Jules Verne in 1878 and Kurd Lasswitz in 1897. In 1895, the space-station concept was noted from a more technical viewpoint in a science-fiction story by Konstantin Tsiolkovsky. In 1903, Tsiolkovsky expanded his description of the manned space station to include rotation to create artificial gravity, use of solar energy, and even a space "greenhouse" with a closed ecological system. Thus, at the turn of the twentieth century, the idea of the space habitat was defined in terms of some of its basic elements. The idea progressed slowly over the next fifty years, then accelerated. In 1923, Hermann Oberth elaborated on potential uses of space stations, noting that they could serve as platforms for scientific research, astronomical observations, and earth-watch. In 1928, Guido von Pirquet considered a system of three stations, one in near orbit, one more distant, and a transit station in an intermediate elliptical orbit to link the other two; he suggested that they might serve as refueling depots for deep space flights. The concept of a rotating wheel-shaped station was introduced in 1929 by Potocnik, writing as Herman Noordung. He called his 30-mile diameter station "Wohnrad" (living wheel) and suggested that it be placed in geosynchronous orbit. During World War II, space stations received some military study in Germany and after the war, the idea surfaced again in technical circles as a geosynchronous rotating-boom concept proposed again by H. E. Ross in 1949.

The space-station idea was then popularized in the United States by Wernher von Braun. In 1952, he updated Noordung's wheel, increased the diameter to 76 miles, and suggested a 1730-km orbit. At about the same time, Arthur C. Clarke published *Islands in the Sky,* a novel involving larger stations. In 1961, in another novel, Clarke suggested placing large stations at the Lagrangian libration points where they would maintain a fixed position relative to both the earth and the moon. In 1956, Darrell Romick advanced a more ambitious proposal—for a cylinder 1 km long and 300 miles in diameter with hemispherical endcaps having a 500-mile diameter rotating disc at one end to be inhabited by 20,000 people.

The companion idea of a nuclear-propelled space ark carrying civilization from a dying solar system toward another star for a new beginning was envisioned in 1918 by Robert Goddard. Possibly concerned about professional criticism, he placed his manuscript in a sealed envelope for posterity and it did not see print for over half a century. In 1929 the concepts of artificial planets and self-contained worlds appeared in the works of J. D. Bernal and Olaf Stapledon, and by 1941 the interstellar ark concept had been fully expanded by Robert A. Heinlein and others, many appearing in the science-fiction publications of Hugo Gernsback and others. In 1952 the concept was outlined in more technical detail by L. R. Shepherd, who envisioned a nuclear-propelled million-ton interstellar colony shaped like a sphere flattened at the ends, which he called a "Noah's Ark."

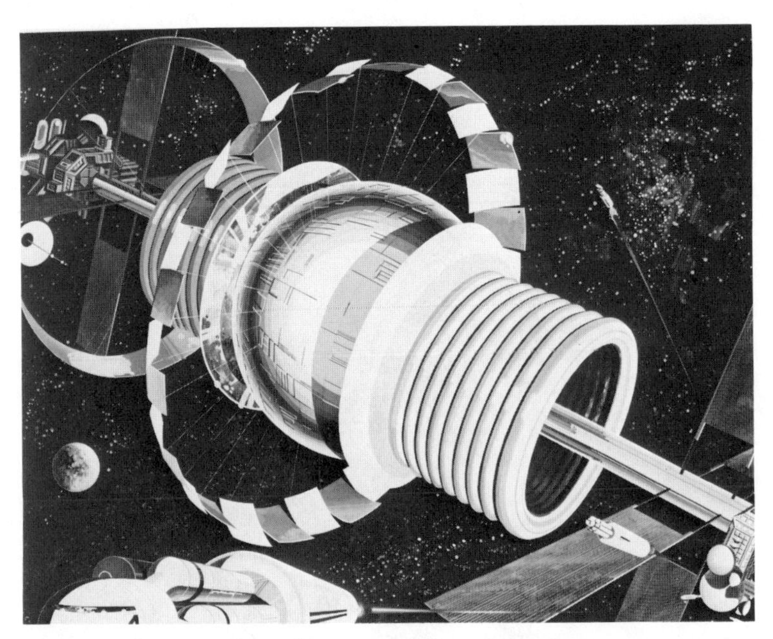

Bernal space colony concept.

Interior illustration of the Bernal space colony.

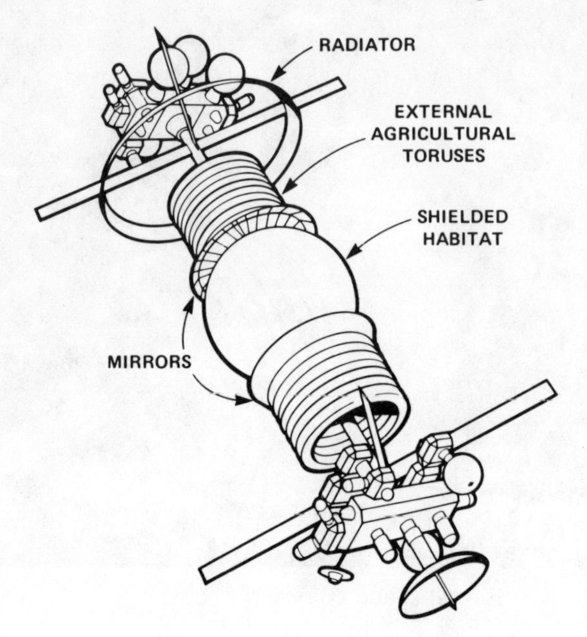

Diagram of the Bernal space colony.

A related idea, the use of extraterrestrial resources to manufacture propellants and structures, was suggested by Goddard in 1920. It became a common theme in science fiction and reappeared in technical literature after World War II. In 1950, Clarke noted the possibility of mining the moon and of launching lunar material to space by an electromagnetic accelerator along a track on its surface. In 1948, Fritz Zwicky suggested use of extraterrestrial resources to reconstruct the entire universe, beginning with making the planets, satellites, and asteroids habitable by changing them intrinsically and changing their positions relative to the sun. A scheme to make Venus habitable by injecting colonies of algae to reduce atmospheric CO_2 concentration was proposed in 1961 by Carl Sagan. In 1963, Dandridge Cole suggested hollowing out an ellipsoidal asteroid about 30 km long, rotating it about the major axis to simulate gravity, reflecting sunlight inside with mirrors, and creating on the inner shell a pastoral setting as a permanent habitat for a colony.

In 1960, Freeman Dyson suggested an ultimate result of such planetary engineering: processing the materials of uninhabited planets and satellites to fashion many habitats in heliocentric orbits. A shell-like accumulation of myriads of such habitats in their orbits has been called a "Dyson sphere."

On July 20, 1969, astronauts Neil A. Armstrong and Edwin E. Aldrin, Jr., walked on the moon. In the context of history just received, the ". . . one small step for a man, one giant leap for mankind" appears quite natural and unsurprising. Most scientists agree, that if the first step is to be followed by others, space colonization may well be those succeeding steps. Perhaps mankind will make the purpose

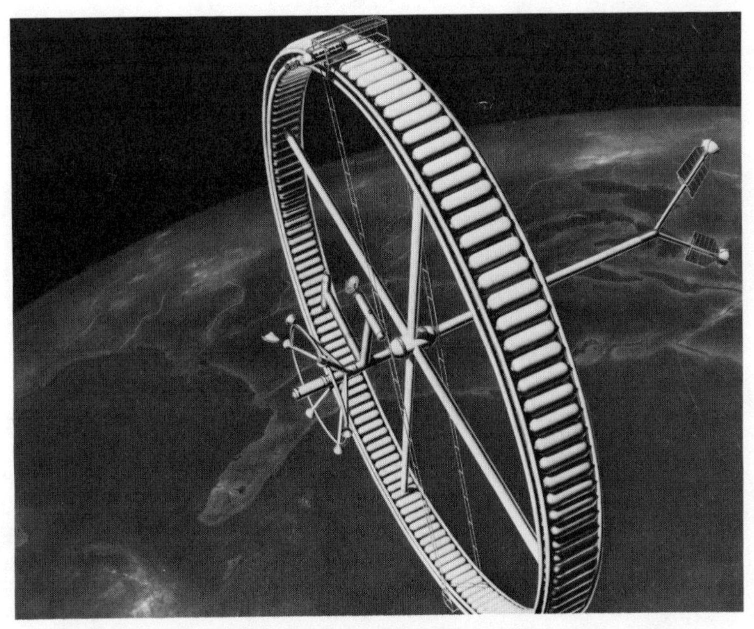

A Lockheed vision of a 1990 space city.

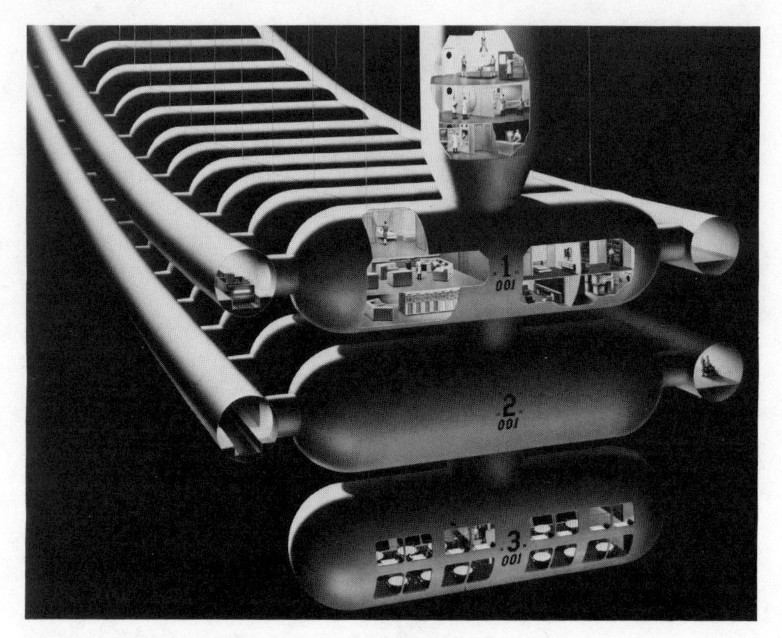

Cutaway of the Lockheed space city.

of the next century in space what Hermann Oberth proposed several decades ago:

> To make available for life every place where life is possible. To make inhabitable all worlds as yet uninhabitable, and all life purposeful.

THE SPACE COLONY: THE NASA/AMES-STANFORD STUDY GROUP

Whether or not America's next major goal in space should be to establish a colony in space with approximately 10,000 people was the subject of study programs hosted by Ames Research Center and co-sponsored by NASA and Stanford in the summer of 1975, 1976, and 1977. Their unanimous conclusion was that while space colonization per se was unable to contribute to a lessening of population pressures in the relevant future, humanity's expansion into space was unavoidable in the long run for sheer survival. There can be little doubt, the scientists argued, that permanent cities in space are in humanity's future, and it is one of man's most important obligations to future generations to keep this and other growth options open.

During the three summer sessions, the study groups designed a system for colonizing space and found no fundamental scientific obstacles to such an undertaking, although the practical engineering and social problems were seen to be quite difficult. Their recommendations called for the construction of a wheel-shaped habitat over a mile in diameter at a point on the moon's orbit 240,000 miles from both the earth and the moon. The torus, the body of the habitat, would rotate around its hub at one revolution per minute, fast enough so that the centrifugal force felt by its

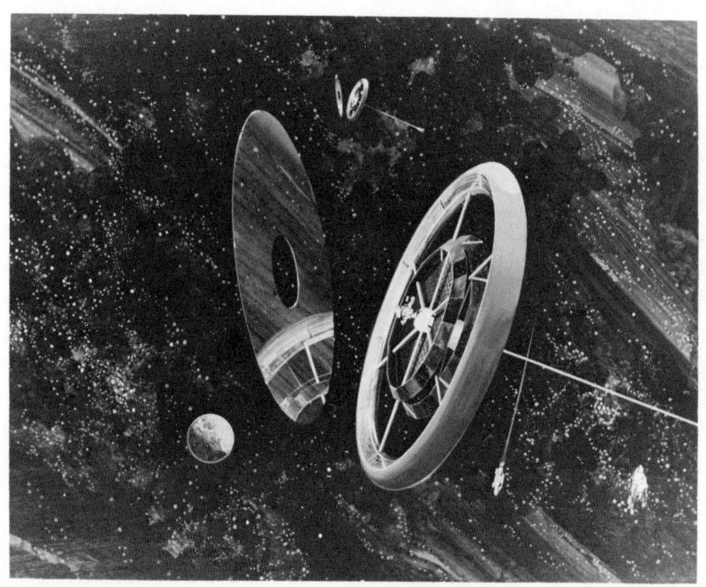

An artist's rendering of a concept developed by NASA/Ames-Stanford study group.

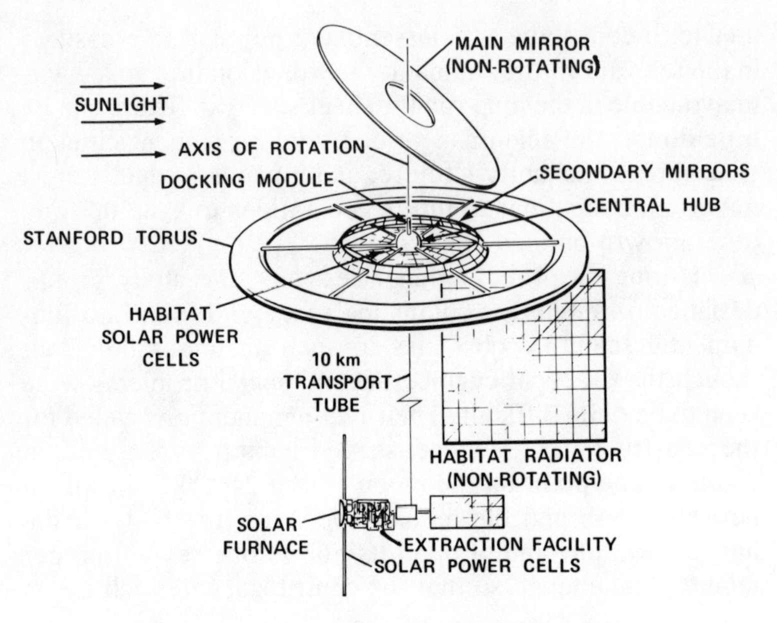

Diagram of the study group's space colony.

inhabitants would simulate their normal weight on earth. Its rim would house 10,000 people along with businesses, gardens, schools, light industry, and closed-loop agriculture. The mass of the torus and its contents would be approximately 500,000 tons, almost the same mass as the largest supertanker. Heavy industry would be outside in order to make use of the weightlessness and high vacuum of space. The major industries would focus on the manufacture of other colonies and on power satellites placed in geosynchronous orbit above the earth. These satellites would gather sunlight almost twenty-four hours a day and beam it to earth as low density microwaves. On earth the microwaves would be converted to electricity and fed into the usual distribution systems.

Using the abundant solar energy of space, space colonies would process materials from the moon. The Apollo explorations have demonstrated that lunar materials contain 20-30 percent metals, 20 percent silicon and 40 percent oxygen by weight. Some 150 engineers on the moon would mine and ship into space a million tons of these materials each year using a ground-based mass launcher powered by a special type of electric motor. The material would be metallurgically refined to extract aluminum (as well as perhaps titanium) for construction material, silica for glass, and oxygen for life support and a rocket fuel. Carbon, nitrogen, and hydrogen would have to be delivered from the earth.

The NASA/Ames-Stanford groups were more conservative in establishing design criteria. The space scientists elected to create the living space as earth-like as possible, especially where the physiological and psychological consequences of deviations from the earth's environment have not been adequately studied. The transport system the study groups selected would use rocket vehicles which are simple extensions of those now under development, namely the shuttle and an easy derivative of it. Structures were designed assuming that they would be built of ordinary, conventional materials such as aluminum and glass, and requiring little or no significant increase in manufacturing productivity of the colony because of automation.

The group chose to simulate the gravity of the earth by rotating the habitat to produce a centrifugal force equal to the gravity on earth. The rotation rate was restricted to one revolution per minute in order to minimize disturbances of the sense of balance. The habitat atmosphere was designed to contain the same amount of oxygen as on earth but with slightly less than half as much nitrogen, so that the total atmospheric pressure would be about half that at sea level on earth.

Away from the earth's magnetic field cosmic radiation is intense. The scientific study groups recommended that radiation exposure be reduced by shielding to a level below 0.5 rem*/year, the U.S. standard for the general popula-

---

*A *rem* is the dosage of ionizing radiation that will affect man physiologically the same as one roentgen of X-ray dosage.

tion. The ten million tons of shielding requiring to reduce exposure in the habitat below this level would come from the slag from processed ores.

Serious psychological problems might develop from living in an entirely man-made structure at high population densities remote from other communities. Therefore, aesthetics were created with the idea of forestalling these problems. The design laid out allows for lines of sight of over half a mile, and the feeling of spaciousness with proximity to growing plants. Special consideration was given to community and architectural planning in order to allow diversity of development and adaptability while also offering the privacy essential in a population density of more than 60 people per acre.

Using the sophisticated techniques on earth to achieve extremely high yields, intensive agriculture would be carried on in segments of the colony. Grains and certain other

Artist's conception of the agricultural area of the study group's space colony.

plants grow faster the more light they obtain, and in the habitat agricultural areas can be provided with sunlight up to twenty-four hours a day. The concentration of carbon dioxide and water vapor in the agricultural areas would be adjusted to maximize growth rates. Cereals, vegetables, poultry, ham, and dairy products for a population of 10,000 could be grown on a total of 160 acres. Wastes from animals, plants, and humans would be turned into pure water and agricultural chemicals by a wet oxidation sewage system like the most advanced type on earth. With fast recycling, minor amounts of water and other essentials will be necessary.

Actual manufacturing and assembly would take place next to the habitat in an enclosed volume containing an atmosphere and gravity adjusted to the most convenient value for the construction task. The components for satellite solar power stations and new colonies would be manufactured in these small factories. Final assembly and testing would occur in space close to the colony, so that engineers would travel within a few minutes between their homes and work.

Another area of study by the Ames study group concerned the social, cultural, and interpersonal aspects of the people living and working in the colony. The analysis assumed an initial population of young, strong, and vigorous men and women in their prime working years, with only a small number of children and older people. Scientists feel that sequence would parallel the history of communities in the American west and other frontier areas during their initial settlements. If space colonization is developed from a U.S. perspective, it's obvious that American cultural habits and conventions will take root. However, if several other nations take part, a wider range of cultural experiences can be expected.

During the early years of space colonization, space scientists expect that the number of habitats will grow rapidly, each colony serving as a construction facility to build others. Rapid social change, an expansion of cultural options, and a diversity in directions of cultural development will therefore characterize space-colonization. Social scientists fear this will be especially pronounced if the space venture is not a single-nation activity. Heterogeneity, mutual enrichment by the interaction of diverse elements, and multiple sources of food, energy, and other needs in a number of closely interacting colonies are seen as positive elements. They should virtually guarantee the survival, endurance, and further development of the cities in their unfamiliar environment.

Another area of concern revolved around the offering of very high salaries to engineers and their construction workers—wages which might not attract the kinds of men and women best suited to long-term space habitation. Instead, it might be preferable to offer certain rights of settlement in combination with subsidies for family support. Socially and culturally, the space colonies will be physically isolated, although the people within would enjoy instantaneous electronic earth communications. Their environment would offer the stimulation which good design offers, attention to personal space, contact with the environment, and freedom to modify the habitat as experience and interest may demand. Evolution toward a considerable degree of self-government within the colonies seems likely as the years go on, which would probably be accompanied by governmental diversity as the number of celestial cities grows.

The physical safety within an enclosed city would be of a serious concern because of the relative vulnerability of the habitat to deliberate damage. Social scientists and psychologists expect the initial small habitats to be less tolerant of dangerous psychotic behavior than are some cultures on earth. Therefore, a selection or screening process will be incorporated which is much less demanding than the one employed for astronauts, although comparable to that of the Peace Corps. A reverse-flow system of people who discover they cannot accept colony life will be instituted.

Scientists expect that during the coming two decades space colonies will be constructed in larger sizes and will be populated less densely as automation reduces their construction cost. Eventually, the asteroids would be exploited for elements of carbon, nitrogen, and hydrogen, elements virtually non-existent on the moon. Space geologists feel the material of the asteroids would be sufficient for the construction of space colonies with a total land area 3,000 times that of earth.

All the NASA/Ames-Stanford scientists agree that the size of individual colonies would eventually be very large. Within the material limits of ordinary civil engineering practice, it is apparent that it will be possible to construct large spherical colonies with diameters of as much as twelve miles (twenty kilometers) in which the total habitable land area would be some 250 square miles (88 km^2). Such small cities built of titanium mined from the moon would be structurally thick enough, having deep enough atmospheres to provide without additional shielding a degree of protection against cosmic rays similar to what is provided for man by the earth's magnetic field and atmosphere.

Thus, the various seminar groups concluded that because space-colonization is possible within the current limits of the technology available to us and may offer real benefits such as an unlimited amount of electrical energy from a clean, non-polluting source, it deserves additional study in much greater depth.

However, on the basis of the conservative requirements on physiology and engineering, the total price of the first city, including all the development work leading toward it, appears to be roughly $100 billion, or between two and three times the cost of the Apollo project in the same 1975 dollars. But in contrast to Apollo, space colonization would be a paying proposition with a benefit/cost ratio greater than one. Thus, it represents "cashing in" on the scientific information returned to man by Apollo. The analysis concluded that there were possibilities for substantial reduction in costs if some of the physiological requirements were relaxed, or if any benefit could be derived from modest technical advances of a kind which were beyond the self-imposed conservative limits.

But there was a very strong belief on the part of many that the numerical figures were less significant than on a qualitative fact: that space colonization appears to present a way out from the sense of closure and of limits which is now oppressive to some people on earth. Growth has been a vehicle of rapid and often progressive social change, particularly in the Americas and other former colonies. Of course, it has been a source of opportunity for millions of people. Earthlings tend to view with distaste a future in which opportunities become increasingly restricted, and in which new and oppressive political institutions would have to be devised in order to equitably allocate resources which were insufficient to meet the demands.

The colonization of space may provide a way to bring new wealth to the earth as well as new opportunities to its people, without the environmental damage which has so often accompanied growth on earth in the past. It may offer a means to permit the continuation of America's industrial revolution, a process which has already brought a considerable degree of physical comfort and freedom of opportunity to most of mankind.

GETTING THERE

Preparation for a trip from earth to low-earth-orbit and eventually to a space colony will be difficult. This is seen as a period to eliminate those who are not serious about traveling to a space colony. Each individual will have to spend weeks in quarantine while undergoing exhaustive physical exams, tough decontamination, and various other analyses to ensure the individual is not carrying insects, bacteria, fungi, etc. Also, the time in isolation will aid social psychologists in determining whether the person has any men-tal disorders. Even slight neuroses would be unwanted aboard the colony.

Only after such complete and stringent tests would the individual be permitted to board a personnel module or a Heavy-lift Launch Vehicle (the HLLV) along with 99 other prospective colonists. From that moment on, events will take place rapidly. The vehicle will be launched and acceleration will thrust the person into a specially designed contour seat. A few minutes later, the acceleration will cease and the individual will find himself 240 km above the earth. His first experience with weightlessness will then begin.

But the orbit is nothing more than a staging area during which time an entire section of the HLLV (the personnel carrier transporting the colonists) is transferred to an Inter-orbital Transport Vehicle (IOTV). This unit is the workhorse transporter which carries people and cargoes between points in space, but never lands upon any planetary body. Its structure will seem frail and delicate compared to the airplane-like structure of the HLLV.

During the construction phase of the colony, the staging area will handle replacement supplies at the rate of 1,000 tons a year. The growth and increasing population loading of the colony will require shipment of an average of fifty to sixty colonists per week, along with their personal belongings and the carbon, nitrogen, and hydrogen required to sustain them in space. But, a while later, the major need will be for lightweight, complex components constructed for solar satellite power stations. The initial resupply of the lunar base will come from earth.

Since the cargo will be carried to the staging area on earlier HLLV flights, passengers will not have to wait long in orbit. This helps reduce the amount of consumables needed to support the people between earth and the colony. Space scientists will make every effort to transport the individual to the colony as quickly as possible after attaining earth orbit. The freight is transferred to the IOTV before arrival so no time will be lost in moving the personnel carrier from the HLLV to the IOTV. The gigantic rocket engines of the IOTV will begin to thrust and the vehicle will break out of earth orbit and commence its 6-day journey to L_5 orbit and the colony.

The colonists will anxiously await their first glimpse of the huge, wheel-like structure spinning in the vast backdrop of black space. As they approach the colony, they will be startled at how small the celestial city appears. Because they will be unable to judge distance in space, the habitat will appear in miniature. As the Inter-orbital Transport Vehicle matches its orbit with that of the colony and prepares to dock, the habitat will appear as a giant wheel in space.

The rough-looking outer "tire" is in reality a radiation shield built of rubble from the nearby moon—the dust and slag protect the people from cosmic rays. This passive shielding against cosmic rays is a separate unconnected shell with a space of approximately 1½ meters between it and the torus. The shield is 1.7 meters thick and built from large blocks of compressed lunar rock held together by special mechanical fasteners. Over the window areas, the shield is designed in the form of chevrons or V-shaped, with mirrored surfaces which transmit light via a series of reflectors while blocking cosmic rays.

A burnished disc hangs suspended above the wheel of the space colony. This is a large mirror which reflects sunlight to outer mirrors, which in turn guide the light rays through several other mirrors arranged in a V-shape to block cosmic rays. The habitat rotates within the outer shield. To simulate earth's normal gravity the entire habitat rotates at one revolution per minute around the central hub.

Walking past the docking module, the colonists notice the walls of the central hub gliding slowly by as they float freely under zero gravity. The colonists have suddenly found themselves in the rotating settlement. But since they are walking near the axis of rotation, the one revolution per minute gives no appreciable sensation of weight.

INSIDE A CELESTIAL CITY

What precisely will this new frontier be like? Imagine for a few moments that you are a visitor on a tour of this new city, strolling among the architecture, agriculture, commerce, and inhabitants. Walk down Main Street, past fountains and fruit trees designed and laid out to relieve the stark simplicity of the manufactured environment; a busy community without freeways and skyscrapers, a beautiful, well-laid out city. The human scale of the city is emphasized by the long lines of sight, the numerous clusters of small gardens, parks, water fountains, and small fruit trees. A sense of openness is created by a broad expanse of yellow sunlight streaming down from high overhead. This area is known as the central plain—running the full circumference of the torus around the middle of the tube. Homes and other living environments are by far the most numerous

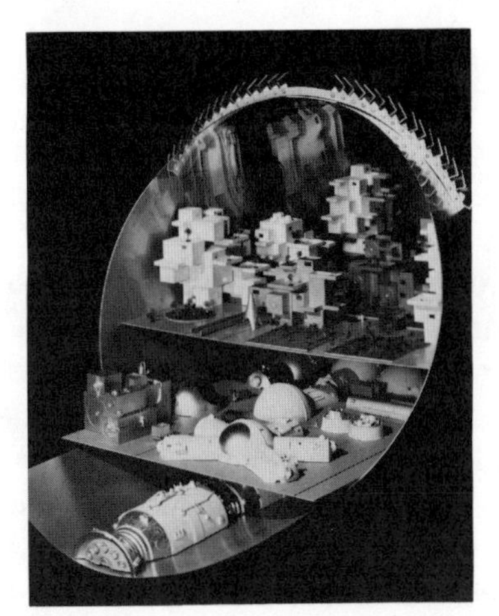

Cross-section of space colony's interior.

Docking station for space colony.

architectural units. Although 106 acres house 10,000 inhabitants, the city nonetheless offers a spacious environment, achieved by terracing the units up the curved walls of the torus. Another way in which spaciousness has been achieved has been to situate commerce (businesses, mechanical subsystems, and light industry) in the portion of the torus which lies below the central plain on which most residents live. The homes have huge window areas in order to further provide the sense of openness. Doors and walls are needed only for acoustical and visual privacy and not protection from the weather or thefts.

The colony has gradually been recruiting people at the rate of some 2,000 per year. Therefore, the homes close to the elevator are already inhabited. Vacant apartments are some 400 meters from where you entered the torus. This is the greatest distance anyone lives from an elevator, and the walk only requires five minutes. Virtually everyone owns a bicycle.

Everywhere there are wide avenues with flourishing vegetation, and little traffic. Stimulated by incredible sunshine, beautiful and brilliantly colored flowers bloom in abundance along patios, stepped terraces, and winding walkways.

A house is a combination of two duplexes and two studio apartments. Each of the studio apartments on the third floor has a small balcony on which some plants are growing. Across the way on a neighbor's balcony is a beautiful stand of strawberries and turnips in several large containers. Small patios below each balcony also have huge pots of miniature fruit trees.

Each individual apartment is comfortably furnished with Scandinavian modern furniture in an attractive and compact manner. However, most of the ornaments and ordinary kitchen furniture are fashioned from plastic, aluminum, and ceramics, a reminder that wood must come from earth or be fused from hydrogen, nitrogen, and carbon brought from earth. Even though the unit has a kitchenette, many lunch and dine in one of the neighborhood's community kitchens. You will be pleasantly surprised to find a meal consisting of steak, green beans, and rice followed by cherry pie and ice cream, hardly the pills science fiction writers have led earthlings to anticipate. Nowhere in the settlement are dehydrated "miracle" foods or algae cake, because the colony is equipped with extraordinarily productive farms which grow food familiar to the earth's inhabitants.

You will notice many young adults, as well as a few children. The settlement's community would primarily consist of men and women between the ages of 18 and 40, plus several hundred youngsters who arrived with their parents from earth, and some 100 children who were born in the colony. Thus, the mix in population is comparable to that of any typical frontier. The inhabitants are hardworking, concentrating all their energies on the manufacture of satellite solar power stations, as well as the construction of the next colony, an exact but much larger replica of the one they are in now. There is an amazing amount of innovation going on despite the colony's limited functions. The gradual but consistent growth of the colony is sustaining a sense of dynamic change. The stabilization of the settlement, upon achieving its full growth, will undoubtedly result in the loss of stimulation and creative innovation. A community as small and isolated as the settlement will probably stagnate and dissipate in both beauty and productivity. Therefore, the development of hundreds more colonies is necessary. Important economically as well as psychologically, the population will become more like that of earth in its age distribution, as time passes. With the productive fraction of the colonists gradually reducing from 70 percent to 30 percent, the amount of production will decrease in time. Only if the total number of people grow rapidly can production in space be kept at its original level and be increased sufficiently to meet the growing demands of the earth's markets for satellite solar power stations. Not only this, the aggregation of habitats into larger communities will enable the settlements to create cultural and technological diversity similar to that which allows the larger metropolitan centers on earth to be focal points of both innovation and disseminators of cultural and technological change.

The settlement is experiencing an egalitarianism of a frontier reinforced by the esprit of a select number of people working together with a sense of mission on a common goal. The inhabitants feel that that spirit, more than any adventuresome challenge, is making the habitat such an exciting place to live. But, the egalitarianism is tempered by certain realistic conditions within the settlement. First of all, the whole colony has a sense of elitism due to the fact that each member was carefully screened as an original colonist. The distinction developing between those with clean and "shirtsleeve" responsibilities and those who work in hazardous, heavy industry, or zero-atmosphere jobs is not developing into any psychological problems, nor

The interior landscape of a 21st century space colony.

is it expected to produce any marked socioeconomic differentiation for at least several decades.

As you continue your walk, you will find yourself in front of hundreds of tiers of fields, ponds, and cascading water. The upper level where you entered is surrounded by a large number of ponds containing over 100,000 fish of different sizes and varieties. From these ponds the water flows down to lower levels where it irrigates fields of corn, sorghum, soy beans, rice, alfalfa and various other kinds of vegetables. The ponds also provide water for livestock. The multiple tiers triple the area of cropland.

On the second tier wheat is growing and should be ready for harvesting in another week. There are three such large areas in the habitat and each grows essentially the same crops. However, the harvests are staggered in order to offer a continuous supply of agricultural growth. On the third tier, huge tomatoes blossom in a specially designed control zone with elevated levels of carbon dioxide, humidity, and temperature. On the fourth and fifth tiers, some 20,000 chickens, 10,000 rabbits, and 500 cattle are caged. The lowest level in the area is enclosed and maintained at an extremely low humidity in order to allow for the rapid drying of crops which in turn will hasten produce flow from harvest to consumption. Due to high productivity, the colony's 160 acres of agriculture provide vegetables for the 10,000 people.

Within the hub off the habitat is a 100-meter diameter fabrication sphere. This is where metals are shaped and formed and where much of the assembly and construction take place. On one side of the fabrication sphere are the solar power plant and furnace used in manufacture. Directly across from the plant is the colony's radiator with an edge facing the sun. This radiator radiates the heat of the colony into space by a complex of heat exchanges passing through the spokes from the torus to the hub. Similar to the habitat's docking area, the fabrication sphere and radiator do not rotate.

You might stop at a small pub for a mug of ale and happen to catch the Oakland Raider and Pittsburgh Steeler National Football League championship playoff game on a huge television screen. But to everyone in the bar, the three dimensional football game played in the central hub is much more thrilling. Only the name of the game played in the settlement is the same since the liberating effects of low gravity make all the passes longer, faster, and more curved, thereby completely changing all the rules and tactics of football.

It is obvious that this new world is so much like earth, yet so completely and superficially man-made. With new lifestyles, visions, and interests for the future, the settlement is nonetheless economically tied to the earth.

Long term economic self-sufficiency and growth require manufacture of products sufficiently useful to earth to attract capital and eventually to create a favorable balance of trade in which the value of exports exceeds that of imports. While super effort is focused on building a number of solar power plants and new settlements, the citizens also hope to minimize imports by producing goods for internal consumption and by maintaining a major recycling industry. Needless to say, the conflict between using resources and manpower for production for internal use and using them for production for export requires a number of creative management decisions. During these initial years of the settlement, the balance appears to be definitely in favor of production for export. Thus, depending on earth as a source of the products and services of highly developed technology as well as for carbon, nitrogen, and hydrogen continues to be significant. On top of this, concentration on exports greatly limits the differences of human initiative in the colony since the majority of productive workers are working in heavy construction. Similar to most of the frontier settlements in early American history, the inhabitants in the L_5 celestial city are primarily concerned with repaying borrowed capital, increasing their standard of living, as well as expanding their base in order to further establish their control over space.

Sanitation Within the Colony

This crucial aspect of the life-support system maintains the delicate balance between the two opposing processes of agricultural production and waste reduction. Back on earth production and waste reduction are balanced (at least partly) by nature's own processes. Water is extracted from the atmosphere by precipitation as rain. Biodegradable materials are reduced by bacterial action. In space, of course, neither of these processes is fast or reliable enough. Without oceans and an extensive atmosphere in which to hold wastes, the settlement is limited in its capacity for biomass and cannot duplicate earth's natural recycling functions.

The colony uses mechanical condensation of atmospheric moisture and chemical oxidation of wastes to reduce the recycling time to 1½ hours. Such a scientific and technical approach minimizes the extra inventory of plants and animals necessary to sustain life and to provide a buffer against malfunctions in the system. Agriculture uses sunlight, carbon dioxide, and chemical nutrients to create vegetation and that is used in turn, to raise animals. Water vapor and oxygen released as by-products regenerate the atmosphere and raise its humidity. Thus, a great amount of vegetable and animal waste is produced along with human wastes of various kinds (sewage, exhaled carbon dioxide, and industrial by-products) which must be recycled. Waste processing also restores to the atmosphere the carbon dioxide used up by the plants, reclaims plant and animal nutrients from the waste materials, and extracts water vapor from the atmosphere to control the humidity of the entire settlement and to obtain pure water for irrigation, drinking, and waste processing. The technicians say that balancing waste generation and waste reduction is a major accomplishment of the designers of the colony, for it eliminates any need to remove excess wastes from the colony thereby avoiding having to replace them with expensive new material from earth.

Water is processed at two main points in the system. Potable water for the people and animals is obtained by condensation from the air. Because evapotranspiration from plants accounts for some 95 percent of the atmospheric moisture, most dehumidifiers are located in the agricultural areas. Thus, because of the rapidity with which plants replace the extracted water, it is crucial that the dehumidification system be reliable. If not, the air would quickly saturate, leading to condensation on cool surfaces, the growth of molds and fungi, and an extremely uncomfortable environment. Several subunits are used for dehumidification.

COMMUTING TO THE MOON

The moon plays a tremendous role in the life of the colony and the settlement of space. For example, the soil in which the various crops are grown comes from the moon. The aluminum used in all the structural elements throughout the habitat is processed from lunar ore. Even lunar rocks have

A lunar mining town showing a transport (foreground), the agricultural bins (with glass windows), the living quarters (three larger tubes), and the magnetic track for launching mined materials (to left of transport).

yielded the oxygen which everyone in the colony breathes. Over a million tons of such ore will be shipped to the colony yearly for construction purposes and to fashion new colonies and the various satellite solar power stations. The mining and transport of the lunar ore from the moon is the primary reason for the successful functioning system of space settlement. The same type of vehicle that carried new settlers from low-earth-orbit to L_5 in five days will take them to the moon.

When the spacecraft has entered lunar parking orbit, it is joined by a small ship known as the LLV (Lunar Landing Vehicle). People then transfer to it through a docking port. After this has taken place, the LLV glides down to the lunar surface and gently settles down by using the retrorockets which create a huge cloud of dust which immediately settles back to the surface in the absence of any atmosphere.

The lunar base offers several services to the engineers and scientists working the two-year shifts of duty. The services include all kinds of recreational facilities, private apartments, and a superb dining complex. All has been designed to make their stay as comfortable and interesting as possible. But, of course, the lunar mining base is a work-base, not a resort area. A monolithic structure fashioned from prefabricated units, it is buried five meters deep in the lunar soil in order to safeguard it from ionizing radiation, thermal fluctuations, and meteorites. The function of the base is to mine, compress, and launch the ore to L_2 (the small space station which serves as a platform to launch exploratory research packages).

In order to supply the one million tons per year of lunar ore to L_5, a surface area the size of some eight football fields must be mined every twelve months. The heavy mining equipment and complex machinery operates over 50 percent of the time, producing a mining rate of almost 4 tons per minute. The ore and soil are dug, scooped, and lifted to a huge processor by scooper-loaders. The huge earth movers shift and mine the moon's plains, digging out the loose rock and dust, minerals from which another world would be fashioned. In nearby control centers, humans manipulate the machine, gouging out huge open pits. The ore is then carried from the mining fields on a long conveyor system to a launch dock where it is compressed into huge launcher buckets and then fused for projection. Such ancient moon-rock is "sintered" or compressed by the sun into 50-pound blocks which form the building blocks of celestial cities.

These lunar blocks would be placed onto carts which would travel at fast speeds down a conveyor belt. Creatively using the moon's weaker gravity and lack of atmosphere, earthlings would literally catapult millions of tons of moon rock to this initial space colony some half million miles away. According to O'Neill, when the cart achieves a velocity of two miles a second, the cart would suddenly stop, shooting the load into space. Every few seconds, another cart would catapult its load into space in a steady stream of huge blocks.

In space, there is an area which scientists refer to as L_5, one of two sausage-shaped limbos in which an object can orbit through eternity. These limbos are delicately balanced in the gravities of moon and earth. L_5, the moon, and earth together form an equilateral triangle. A habitat placed in orbit within L_5 could circle the earth forever. Equidistant between L_5 and the moon, a space net would catch the lunar bricks in flight and channel them to a special construction module already in orbit at L_5. A work crew of 3,000 highly trained engineers would spend two years pumping life into the inorganic moon rock and dust. Meanwhile, an additional team would be busy constructing the new space city. Similar to a huge spider spinning a web the colony would spawn another city. Each, and later others, would be entirely sufficient, non-dependent upon the nearby earth. O'Neill believes that some day soon such self-supporting enclaves would be as numerous as stars. Says the physicist, "Tens of thousands of glowing cylinders will be spinning

Lunar supply vehicle.

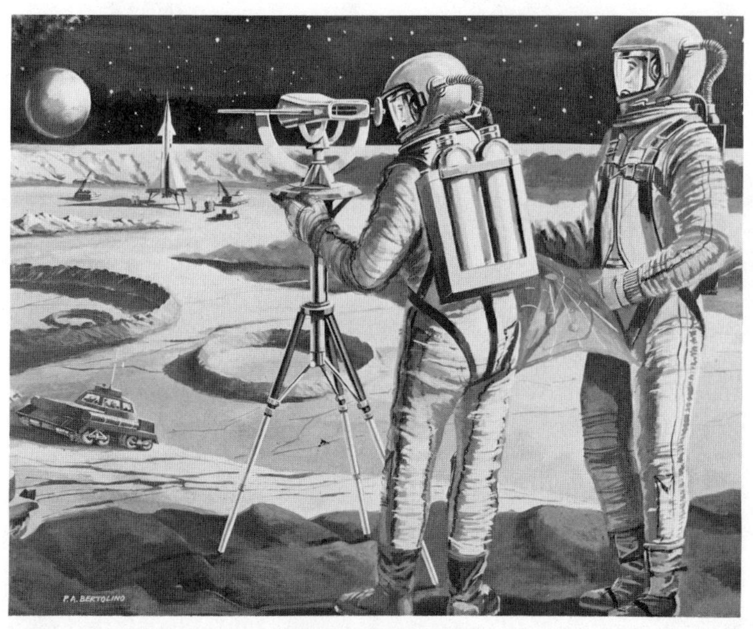

Lunar operations.

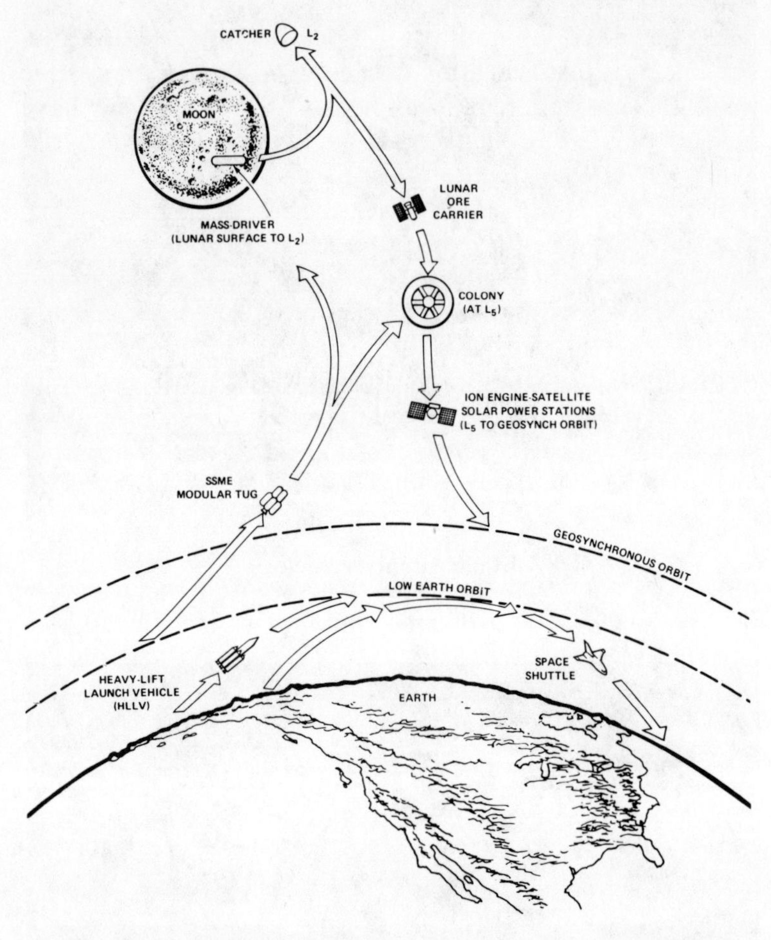

Diagram of space colony transportation system.

around the earth, billions of people living in space and our small and overburdened planet will have a chance to recover."

During the two months the new colonists have been away from the earth, they have traveled over 750,000 km. Now, they prepare themselves for a visit home, an additional 386,000 km. Having lived in a tiny settlement of 10,000 men and women, and having crowded into small bases on the moon and at L_2, homesickness has been inevitable. As planned after two months of debriefings and work, they prepare to leave the colony for earth. They have been fortunate to obtain a berth in one of the shuttle crafts which carries supplies to the moon and rotates personnel from the moon base directly back to earth. The group learns that during the early years all the men and women of the base went straight back to earth and so the personnel transporter was always full to capacity. Now, however, most of the colonists have decided to spend their rotation time at L_5 instead of on earth and berths for the return trip are available. On board during the gentle ride to earth, the group settles back into their comfortable seats reading various terrestrial news magazines and discussing the future of the colonization of the solar system.

Space scientists insist there are no insurmountable problems to prevent human beings from living in space. Obviously there are problems, and they can be solved with currently available technology. It is clear that the people of earth have the knowledge, resources, and courage to colonize space.

V

Strange New Worlds: Interplanetary Exploration and the Search for Intelligent Life

O dark dark dark. They all go into the dark, The vacant intersteller spaces, the vacant into the vacant.

T.S. Eliot, "East Coker"

If only one planet, our earth, circled the sun, and if the earth had no moon, space travel would be an impractical dream. Lonely in the light years of emptiness, colonists of courage would have no convenient intermediate astronautical targets to land upon. But in reality, space explorers have stepping stones to the edge of the solar system, and these planets are so unusual and different from the earth that man's innate curiosity demands that he explore them. Actually, the basic questions that man has had about the solar system are very old and very broad: What are the planets like? Do they harbor life? How did the solar system originate and evolve? Such questions date back to ancient times when man first began to realize that the earth was part of a solar system in which other planets existed. From Babylon

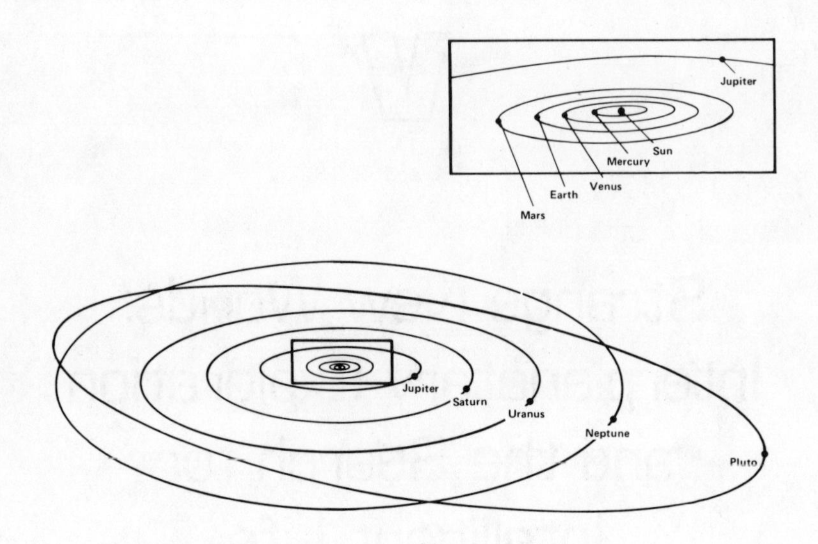

Solar system comparison of planetary orbits.

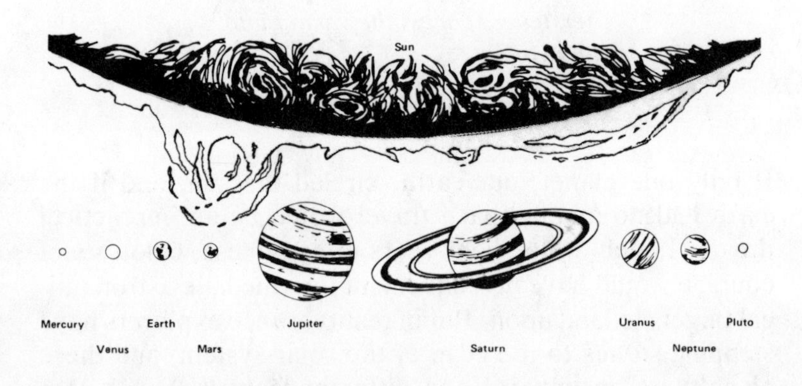

Comparison of planetary sizes.

Comparison of planetary characteristics.

	Mercury	Venus	Earth	Mars	Jupiter	Saturn	Uranus	Neptune	Pluto
Mean Diameter, miles	3010	7650	7926	4240	86,900	71,600	29,400	26,800	3560
, km	4842	12,322	12,742	6664	139,785	115,064	47,402	43,070	5734
, Earth = 1	0.38	0.967	1.000	0.532	10.97	9.03	3.72	3.38	0.45
Volume, Earth = 1	0.06	0.86	1.000	0.15	1318.0	769.0	50.0	59.0	1.1?
Mass, Earth = 1	0.0543	0.8136	1.0000	0.1069	318.35	95.3	14.58	17.26	<0.1?
Mean Density, H_2O=1	5.46	5.1	5.517	3.97	1.334	0.684	1.60	2.25	<5.5?
Oblateness, Flattening of Poles	0.0	0.0	0.0034	0.0052	0.062	0.096	0.06	0.02	?
Escape Velocity, fps (mph)	13,800 (9000)	33,800 (23,400)	36,700 (25,200)	16,400 (10,800)	200,000 (133,200)	121,000 (82,800)	72,200 (46,800)	82,000 (50,000)	<17,300 (11,800)
Surface Gravity, Earth = 1	0.83	0.87	1.00	0.39	2.65	1.17	1.05	1.23	<0.5?
Mean Dist from Sun, million mi	36.0	67.2	93.0	141.5	484	887	1785	2797	3670
Closest Dist to Sun, million mi	28.6	66.8	91.4	128.4	460	837	1700	2775	2761
Max Dist to Sun, million mi	43.4	67.7	94.6	154.9	507	936.5	1860	2820	4610
Mean Orbital Velocity, mi/hr	107,000	78,400	66,600	53,900	29,200	21,580	15,200	12,140	10,600
Period of Axial Rotation, days/hrs/min/sec	179 Days	243 Days (rotates clockwise)	0/23/56/4.1 1 Day	0/24/37/22.7 1.88 Day	0/9/50/30	0/10/14/24	0/10/49/?	0/15/48/?	6/9/?/?
Lowest Visible Surface	Solid	Cloud	Solid	Solid	Cloud	Cloud	Cloud	Cloud	?
Max Visible Surface Temp, °K	611	307	295	260	128	125	103	108	132
, °F	750	210	140	90	-200	-240	-270	-330	-370
Number of Known Satellites	0	0	1	2	13	10	5	2	0

to Alexandria, astronomer-priests focused their attentions upon the seven points of light wandering across the field of fixed stars. The so-called "Sacred Seven" led to the establishment of seven as a mystical number to the ancients. But the "seven" became a very unmystical eight when Neptune was discovered through the telescope by J. G. Galle in 1846. Today, astronomers know that there are at least nine planets, with the slight chance that one or two additional small ones are swinging slowly undetected around the sun along the fringes of the solar system. Planet discovery then grew into a passion with astronomers during the 1800s and early into the 20th century. When Percival Lowell's *Mars and Its Canals* was published in 1906, the whole world became agog about the possibilities of intelligent life in nearby space. Where Ptolemy, Tycho, Bralte, Kepler, and other astronomical pioneers devoted their lives to describing planetary motion accurately, today's space scientists are targeting their efforts toward a more complex strategy: interplanetary exploration to discover extraterrestrial life.

But even though a great deal has been learned over the past three centuries, most basic questions remain unanswered. Man's recently developed technology to land and scan other planets has further whet his appetite to discover more about the solar system, offering hope of answering some of the long standing questions within the next decade. During the past ten years, however, knowledge of the solar system was restricted to a simple understanding of its physical and functional characteristics, such as the positions and orbits of the planets, their density and size. Most of this information has been available for nearly a century. During recent years, ground observations have offered additional information about the chemical and other physical characteristics of the planets and their atmospheres. Yet, all this data is still not enough to develop and test adequate theories on the nature and evolution of the solar system. Today, there are many theories, but most contradict each other. Astronomers and space scientists need more information on the chemistry, physics, age, and evolutionary history of other bodies in the solar system.

At long last, man's technological ability to explore space in depth has offered the means to obtain the wished-for information. The initial step has been to obtain the chemical composition and temperature data on the inner solar system. Space geologists have also began to learn about the evolution and age of the moon from the samples of lunar ore returned. Looking ahead to the 1980s, scientists are already hard at work planning the exploration of the outer solar system and the search for life elsewhere. The ability to search for life on other planets, everyone agrees, is a significant new dimension of solar system exploration.

According to NASA's space officials, the "where, how, and when" of life perplexes thinking people, but we have began to piece together some of the puzzle's pieces. One crucial step has been the combination of the biological and chemical sciences in the study of earth life. Scientists realize that life forms have had a major impact on the evolution of the earth during the time of the existence of the life forms. But through space exploration, scientists are now able to search for life. And, if it is discovered, they can analyze the role that particular form of life has played in the other planets' evolution. Such information will aid scientists in creating a total picture of how other bodies in the solar system have evolved and to better understand the current stage of their evolution. Such technical comparisons of the planets with each other, as well as with the earth, will provide the breakthrough to learning significant information about ourselves and the origins of our solar system.

The exploration of the solar system is not just a matter of visiting other planets in the solar system to simply see what they are like. Instead, it is a systematic search for answers to key scientific questions which will enable scientists to test the more credible theories, or allow them to formulate new ones.

LOOKING FOR EXTRATERRESTRIAL LIFE

From the studies of life on earth, scientists know that all forms of life, from microbe to man, are made from only a handful of the multitude of possible organic molecules. Most crucial among these building blocks of life are sugars, fats, proteins, and nucleic acids. Scientists also know that life forms have developed in complexity and ability to influence the evolution of the planet during the 3½ to 4 billion years that life has existed on earth. In their highly technical laboratories, scientists have been able to synthesize the essential molecules of which life is composed by recreating the primitive atmosphere of the earth and applying one or more forms of energy known to exist on the early planet—energy forms such as solar radiation, electrical discharges, or heat. The ease with which scientists are able to synthesize the organic molecules under simulated prebiological conditions is remarkable. Actually, it suggests the probability that life also originated on any other planet which has undergone a similar evolution. Scientists believe that chemical evolution, the production of those chemical precursors to life, might have occurred in other places in the solar system.

Scientists have already learned a great deal in terms of the existence of prebiological chemicals in space. Meteorites which have come from the asteroid belt have been dissected chemically and found to house many of the organic molecules synthesized in the prebiological experiments on earth. Such analyses corroborate the theory that prebiological organic chemical synthesis is a widespread phenomenon, and the belief that life may well exist elsewhere in the solar system, as well as in other places in the universe.

Among the planets that are accessible, Mars was always considered the most likely to harbor life, although Jupiter is of considerable interest, as is Titan, the largest moon of Saturn. Mars was at first considered promising because of its temperate atmosphere and the knowledge, based on laboratory experiments, that some biological life forms could survive in the Mars environment. Jupiter and Titan have always been of interest because they possess atmospheres of methane, one of the components of the earth's early atmosphere. Thus, it is possible that organic material could survive and accumulate as it was synthesized in the atmosphere.

Scientists are now able to at least frame the questions relating to life elsewhere in the solar system. The foremost question is whether life exists today or has ever existed elsewhere in the solar system. If life is ever found, scientists will want to discover the nature of its basic chemistry and whether it is composed of the same organic molecules as life on earth. On the other hand, if it is learned that no life currently exists on other planets, then the question will be: How far has chemical evolution progressed toward life? Along this line, scientists will try to learn whether the organic molecules, those building blocks of life, were synthesized without actually leading to life, and if so, why. Of course, scientists will not be able to predict all the implications of the answers to these questions. Actually it will take several decades of intensive exploration to answer all the questions adequately. But such answers, as well as insights regarding life elsewhere, of the chemical and geological evolution of other planets, plus the interaction of their biological, chemical, and geological evolution will have an impact on how earthlings treat their own planet in the future in order to maintain it as a place where life can continue to flourish.

During previous life-probing explorations, most of man's information was obtained employing highly complex methods. Initial projects using spacecraft to explore the solar system have been focused upon the moon, as well as the two nearest terrestrial planets, Venus and Mars. Because these are nearest to earth, technical requirements of booster capability, spacecraft lifetime, and communications are far less stringent than the requirements for exploration of the more distant planets. Their closer proximity to earth has also allowed for intensive ground-based observations which in turn has yielded more information about the moon, Venus, and Mars than any other planets in the solar system. Thus, scientific questions about each of these bodies can be more clearly articulated and spacecraft instruments more confidently designed and engineered.

THE MOON

The Project Apollo operations ended on December 17, 1972, with the splashdown and recovery of Apollo 17, "America," in the South Pacific. A dozen men had strolled and rode over 60 miles (110 kilometers) on the surface of the moon, spending a total of 160 man-hours and conducting more than 50 major scientific experiments.

What did they learn?

First of all, the Apollo data reveals the moon is a planet which formed over 4½ billion years ago, the approximate time the solar system came into being. It is composed of a large proportion of the high-temperature, early-condensing solar nebula compounds. During the first few hundred million years of its existence it formed an aluminum-rich highlands crust and iron-rich mantle, both depleted in volatile compounds such as water and methane. That early time was one during which the moon was being heavily bombarded by asteroid and meteoroid impacts, the larger of which shaped the great circular mare basins. Somewhere between ½ to 1½ billion years after lunar origin, the frontside mare basins (the dark areas of the moon's surface) located in a thinner crust than far-side basins filled with basaltic lava which originated from melting in the deep iron-rich interior. Since that time, the moon has been dormant in a volcanic sense. Today, its seismic energy release is almost one billion times less than that of earth. The lunar peace has really been to man's benefit since it has provided

Astronaut David Scott on the lunar rover, August–September 1971.

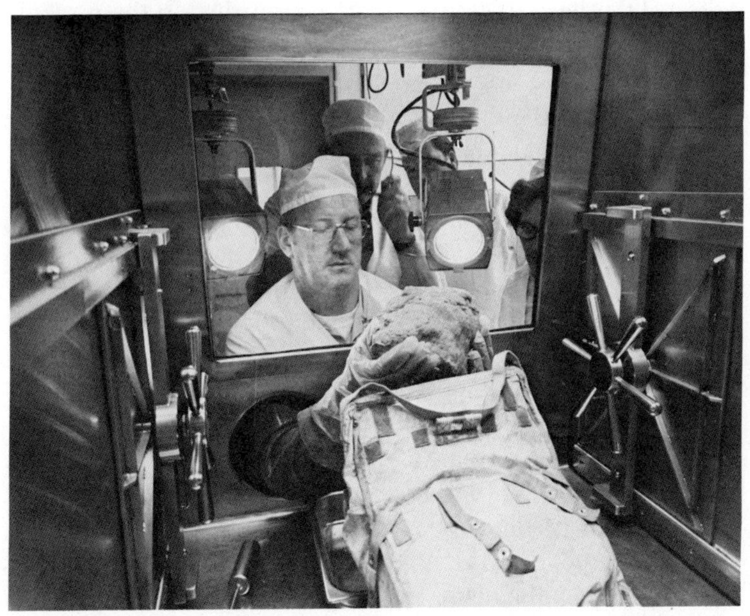

"Big Mulie," rock collected from the moon by Apollo 16 crew.

us an accessible surface which has recorded the history of meteoroid, comet, solar wind, and cosmic-ray activity for over 3 billion years.

Apparently, the moon's mass is on the side nearer the earth and the inner-most core may still be very, very hot. The moon's chemical composition is very much like the earth's, but in different proportions.

Today, the findings are steadily streaming in from laboratories in sixteen nations where scientists have been analyzing the 841 pounds (382 kilograms) of rock and soil, plus the 9,000 photographs brought home by Apollo. After the last mission was flown, a dozen experiments planted on the moon in five lunar laboratories have continued to send data back to earth. Instruments set up by the Apollo expeditions have detected distinct layering beneath the moon's surface. The moon appears to be solid to a depth of some 500 miles (800 kilometers), which is almost halfway to the center. The moon is at least partially molten at depths below 600 miles (1000 kilometers). Contrary to the prevailing pre-Apollo opinion that the moon was cold, the moon appears to have a warm interior. Instruments have detected heat flowing outward from the moon's interior at a rate slightly lower than that of earth but surprisingly large for the moon's size.

The lunar story is a fascinating story in its own right, more so in terms of its applicability to understanding how the earth and other planets formed and evolved. Scientists feel that if a planet as small as the moon formed a crust soon after formation, so probably did all planets of similar or larger size such as Mars and earth. In fact, geologists now speculate that some 2½ to 3¾ billion-year-old aluminum-rich rocks found on earth are remnants of an early crust, one which possibly formed in the same manner as the lunar highlands. It further appears that other old earth volcanic rocks may be equivalent to lunar mare basalts, perhaps representing volcanism triggered by giant impacts now known to have occurred on the earth 4 billion years ago.

The analysis of lunar soil samples leads scientists to believe that solar activity has been relatively constant over the past several million years. This may shatter the belief that the earth's glacial cycles, or ice ages, are related to changes in the sun's activity. Embedded within the soil samples were solar wind atoms over ½ billion years old which appear to have the same composition as today's solar wind. Such atoms are also found in some 4-billion-year-old rocks from the lunar highlands. Astronomers and space scientists expect their analyses over the next decade to provide them with a reliable history of much of the sun's activity since it formed, a record of understanding the sun's evolution. Scientists also suspect that the very low amount of carbon, nitrogen, and water on the moon precluded formation of bio-organic compounds. This result reinforces the belief that life-detection missions have to be directed toward the volatile-rich planets.

MARS

Of all the magnificent splendors in the night sky, few have caught man's attention, or impelled his imagination, more than the planet Mars. Writing in his *Swords of Mars,* published in 1934, novelist Edgar Rice Burroughs wrote, "It is a long story, a story of love and loyalty, of hate and crime, a story of dripping swords, of strange places, and strange people upon a strange world . . ."

How long ago others started believing that strange people inhabited the red glittering point slowly moving across the heavens is lost in the shadow years of prehistory. But one fact is certain: the red dot glowed with a steady, reddish hue, very different from the hard, brilliant pinpoints of white light from the stars. And, where those stars appeared to glide through the various seasons in an orderly march, Mars followed its own tune, first drawing close to earth, then receding. The unusual color and wandering led the ancients to believe that it was governed by special supernatural powers. To the Babylonians, for example, Mars was *Nergol,* the god of life, a bold warrior. From this early concept, the Greeks and Romans attributed omens of battle and slaughter to the red body in space. *Mars* is the name for the Roman god of war.

Down through the past 2,000 years, the planet has continued to intrigue the earthlings. With the invention of the telescope, which allowed astronomers to observe something more of the dot than its color, the human mind projected onto the planet the full vent of imagination. The indistinct patches and surface features were sharpened and infused with meaning and purpose more in line with the experiences known on earth.

With J. H. von Mälder's topographic map of the surface of Mars appearing around 1840, the planet became the object of intensive study for nearly every astronomer and amateur with a telescope. Maps of the planet appeared regularly as telescopes became more powerful and different scientists added bits and fragments of new information. The Proctor map of 1867 showed bright reddish zones, which were believed to be continents, and dark bluish zones which were thought to be seas.

In 1867, physicists Pierre Jules Janssen and Sir William Huggings made the first attempt to discover oxygen and water vapor in the Martian atmosphere. Without oxygen or water vapor, life in any form would be impossible. The first attempt to find these crucial substances failed. Tests yielded neither positive nor negative results. However, the experiments of the two young physicists demonstrates the kind of interest that Mars had on scientists.

The idea of Mars as a planet of seas and oceans was short lived. As early as 1863, Giovanni Schiaparelli, the director of the Milan observatory in Italy, remarked that the dark areas of Mars did not reflect the sun's image as sheets of water should. Furthermore, Sciaparelli dumbfounded as well as fired the imaginations of the entire scientific and literary worlds with his discovery of "canals" on Mars. He added, "The canals cannot be anything but the work of intelligent beings. This is nothing impossible."

Fourteen years later in 1877, the canals were recorded in detail by Schiaparelli. The "canoli" (Italian for channels), insisted the astronomer, cover the vast areas of the continents. "They traverse the planet for long distances in regular lines that do not at all resemble the winding courses of our streams. Some are easy to see; others are extremely difficult and resemble the finest thread of a spider's web drawn across the disk."

From 1877 to 1886, Schiaparelli was considered more a science fiction writer than a credible astronomer. No other scientist believed his descriptions of the canal system because no telescopes but his seemed to show them. Furthermore, his telescope was very small compared to others across the industrialized nations. But, in 1886, Thollon and Perrotin employing a 30-inch telescope in their Nice, France, observatory saw Schiaparelli's intertwined network of canals. Within a few years, astronomers in other western European nations began recording the canals on maps. In 1890, Schiaparelli retired from observing because of his poor eyesight. William H. Pickering and Percival Lowell took up the cause and origin of the canals. Over the next three decades Lowell was to become the prime authority on Mars and its possible life. In fact, he became so convinced that Mars was inhabited by intelligent beings who had constructed the canals that he coined the name "Martians."

A brilliant mathematician from Boston, Lowell predicted the probable position for the unseen planet Pluto from minor disturbances in the Neptune orbit. Later, he founded his own observatory at Flagstaff, Arizona, where he observed the solar system through the clear desert air. He equipped his observatory with the most advanced and precise instruments. Pluto was finally discovered and observed for the first time from this observatory based upon his most advanced calculations. From 1895 on, observations of Mars by the thousands were made by Lowell. He told the public, "There are celestial sights more dazzling, spectacles that inspire more awe, but to the thoughtful observor who is privileged to see them well, there is nothing in the sky so profoundly impressive as the canals of Mars." Later, in his *Mars and Its Canals*, Lowell created a public legacy by writing, "That Mars is inhabited, we may consider as certain although it is uncertain what those beings may be." Yet, he was privately convinced that what Schiaparelli had observed and he himself had expanded upon was a huge irrigation system which Martian engineers had designed in order to keep their dying planet green and fertile. In his book *Riddle of Mars*, C. E. Housden laid out the form and the power of the supposed pumping stations needed to irrigate Mars as the Martians appeared to be doing.

Thus, the red planet was catapulted for good into the public's imagination as a strange new world teeming with life. Although the ancients were the ones who provided 20th century man with the rudiments of a philosophy of science with which to investigate the solar system, and although it was the great minds of later centuries who provided the observational instruments, it was really the later decades of the 19th century which contributed the most to the concepts of Mars seen in literature: a planet of canals, and oases—a very dry desert with ditches and canals and conservation on a vast scale required for the existence of life. It was a planet inhabited by war-like intelligence, evil deeds, and with science beyond the dreams of earthlings, or philosophies beyond imagination.ts of human imagination.

To many early astronomers, it seemed certain that Mars was inhabited, just like the earth. And, if its creatures and plants were bizarrely depicted, it was because Mars was populated with the rich endowments of human imagination.

And now, some seventy years later, human intelligence has succeeded in populating the planet with its most incredible achievement yet: a Viking robot. Its "looks" are perhaps as bizarre as any Martian creature ever conceived by a science fiction author. The sides and top of the spacecraft's six-sided box structure are covered with strangely shaped funnels, tanks, covers, pedestals, booms, and antennae. This package of scientific equipment is a triumph of function over form. It was designed by NASA and Martin Marietta Aerospace engineers not as some streamlined space cruiser, but rather as a sturdy landing barge, a craft which could land and position itself safely upon the rugged Martian landscape. Once there, the plan was to carry out a wide range of scientific investigations. On July 20, 1976, the Viking spacecraft began doing exactly that.

> The cameras were the first instruments to reveal something about Mars. No sooner had Viking and its precious cargo of experiments settled on the surface than its on-board computer ordered one of the two cameras to take a picture of the immediate landing area. . . .
>
> It was indeed a remarkable picture, despite the fact that it was in black-and-white and despite the fact that it revealed an area not much bigger than what a person would see hanging his or her head out a car window and staring straight down. There were many small rocks in the Martian scene and a soil as seemingly hard-packed as wet sand. And then one of the Viking lander's footpads came into view—a circular metal dish that had stamped Mars with humanity's mark as indelibly as astronaut Neil Armstrong's boot stamped the moon seven years earlier on this very date.
>
> Minutes later, the camera had lifted its gaze and taken a sweeping panorama of the landing area. . . .
>
> The panoramic picture disclosed that Viking 1 had come down on a rolling, rocky sector of the Martian basin called Chryse Planitia. Parts of the lander intruded in the picture, but rather than detract from the 300-degree breadth of the picture, they gave a reassur-

Mars as photographed by Viking 1 on June 18, 1976.

Mars as photographed by Viking 1 on June 17, 1976.

Viking Lander mockup.

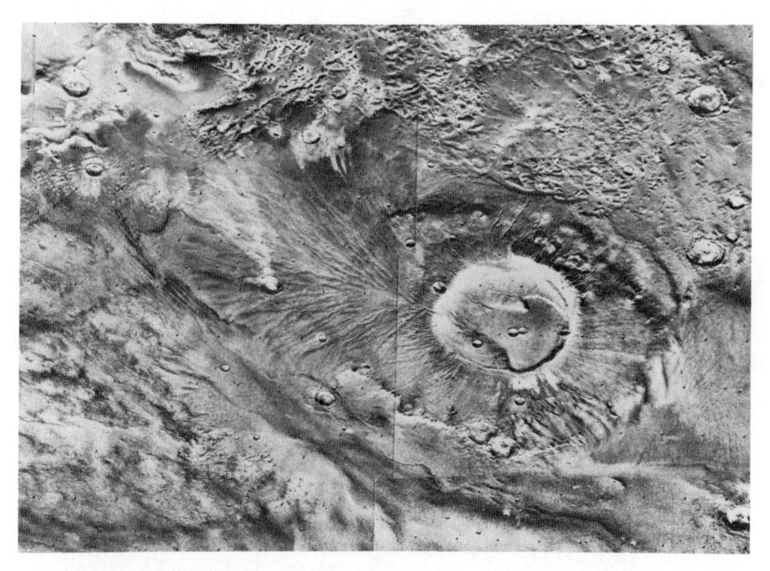

Martian volcano Apollinaris Patera photographed by Viking 1 from a range of 7,400 kilometers (4,600 miles), September 18, 1976.

> ingly familiar sense of scale and orientation to the image, as if the human viewer were actually gazing out on the scene from the windowed bridge of a ship.*

In the weeks and months that followed, the data which poured in from the inhuman little robot's signals offered the world new insights about the mysterious planet. New information revealed that Mars is a much more dynamic planet than was expected. It is clear that volcanic activity has taken place on the planet. Also, its surface, pockmarked from prehistoric crater impacts, is "moonlike" in a variety of ways. Of major interest is a giant canyon slicing through the equatorial zone for 2,300 miles (4,000 km). Burrowing to a depth of over 20,000 feet (6,500 meters), the canyon stretches some 150 miles (250 km) wide.

What could have caused such a huge canyon to form? How were its apparent tributaries created? Were they eroded by water which once existed in abundance on the planet? If this is the case, then Mars must have had a very dense atmosphere. What possibly could have caused Mars to lose its atmosphere?

In order to find the answers to these questions and allow for a clearer understanding of the geological and chemical evolution of Mars, and therefore the earth, earthlings packaged exquisite and delicate instruments into a 1,300-pound laboratory to probe and dig and smell and photograph. All the data is then analyzed within the observation lander and communicated to earth. That incredible laboratory/lander was baked for two days at 233°F to rid it of earthly microbes. With an Orbiter spacemate, Viking was launched safely and accurately, guided for nearly a year over a path of more than 400 million miles to its target on Mars. Once over Mars, Viking had to find a safe landing

---

*George Alexander in "Tantalizing Viking Scientists: Viking Science" in Martin Marietta's *Today Magazine*, No. 3, 1976, p. 7.

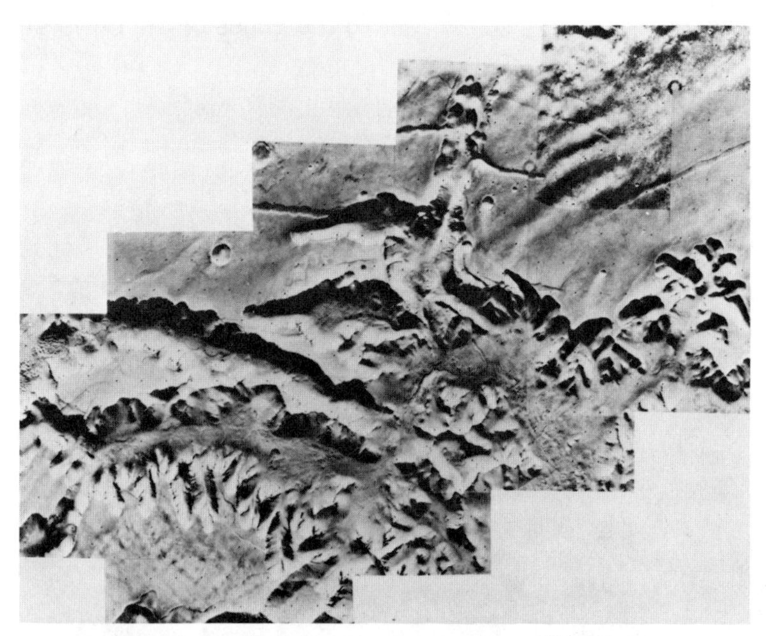

Mosaic of Martian Landscape from Viking 1.

Martian landscape showing dunes and rock fields, Viking 1, August 1976.

Martian landscape, Viking 2, September 1976.

spot and then maneuver with extraordinary precision and an intricate set of descents which placed it down gently. Within moments, it was apparent that the magnitude of the scientific success would reveal itself fully over a period of years. The quantity of information sent back since that 5:12 AM touchdown on July 20, 1976, has already been enormous.

Mars is the fourth most dense planet after the earth, which is the most dense. Pluto is so distant that measurements are difficult and it is uncertain whether it is more or less dense than earth. The difference in specific gravity between earth and Mars is 1.55 (water being equal to 1.0). One cubic centimeter of terrestrial material has a mass of 5.52 grams as opposed to the same volume of Martian material which has a mass of 3.97 grams. Mars' surface density is apparently not uniform. Saturn, on the other hand, has the least density of the planets—less than 1, which means it would float on water. Also, Mars is the eighth-smallest planet in mass. Only Mercury is less massive. If earth mass is equal to 1, Mars is equal to 0.11 and Mercury would equal 0.06. In volume, Mars is only eighth among the planets, having approximately 15 percent of the earth's volume. Mean gravity on Mars is about 40 percent that of earth. For example, a 150-pound man on earth would weigh about 60 pounds on Mars. Yet, there are gravitational variations on the planet which are as great as 17 times more than any measured on earth. Apparently, it has no extensive magnetic field and no radiation belts like the Van Allen belts around the earth. Data from the Soviet Mission Mars 5 confirmed the existence of a magnetic field 7 to 10 times greater than that of space, although the planet's magnetic field is less than 1 percent of the strength of earth's. Thus, Mars probably does not have an iron core at its base.

Similar to earth, Mars is an oblate spheroid. That is, the two poles are slightly flattened by the centrifugal force generated by the planetary rotation. Most all the large planets show significant flattening at the poles, except Venus and Mercury which show none. Pluto is too distant for flattening to be detected. But for practical purposes such as navigation, the planets with little or no flattening (such as Mars and Earth), can be considered spheres.

Phobos and Deimos, the two Martian satellites, were discovered in 1877. Both are very small and revolve close to the surface of the planet. Phobos revolves faster than Mars rotates. It is the only satellite in the solar system known to behave in this manner. The orbits of the two satellites are nearly circular and lie close to the planet's equatorial plane. Until recently, almost nothing was known about the physical appearance of the two satellites, although Phobos was the better understood of the two because its shape and size can be estimated from photographs. Phobos is some 12 miles across and elongated in the place of its orbit. It appears to have an extremely dark surface. Both satellites are synchronous (that is, each turns on its own axis during one revolution around the planet so that the same side always faces the planet).

As far as Mars' surface color is concerned, scientists believe that hydrated iron oxide is the cause of the pinkish-ocher color of the planet's bright areas. These occur primarily in the northern hemisphere. Dark areas appear primarily in the southern hemisphere. Light and dark areas appear to be unrelated to surface altitude.

Mars has a thin and nearly cloudless atmosphere. Near the equator the average surface temperature varies from 314°K (25°F) in early afternoon to 180°K (−135°F) just prior to sunup. This low temperature results in a thin layer of frozen carbon dioxide over the region near each of the poles.

Much of the surface of Mars is covered by a mantle of powdery dust. During occasional periods when the winds reach velocities of over 125 miles per hour, the dust is whipped high up into the atmosphere creating colossal dust storms. The triggering mechanism of these storms is not understood, although it may well be related to disturbances in the atmosphere resulting from heating by the sun. Such a huge storm literally encompassed the entire planet when Mariner 9 approached Mars in 1971. As scientists watched from the satellite's vantage point in orbit, they noticed the changes in thermal structure of the atmosphere as the storm gradually cleared, once again allowing the sun's rays to penetrate the atmosphere to the surface. Such data is helping them to understand the sun's heating and the circulation of the earth's atmosphere. But what the dynamics of Mars mean in terms of the existence of life on that planet is still very uncertain. However, the indication that volcanic activity has taken place there has renewed the interest somewhat dampened by the Mariner 4 discovery of a heavily cratered "moon-like" surface on Mars.

VENUS

Until less than a decade ago, space scientists believed that some forms of life existed beneath the thick clouds completely enshrouding earth's sister planet, Venus. Some even envisioned a tropical environment of flourishing flora and fauna. However, a completely different picture of Venus has emerged based upon radio astronomy observations and unmanned space flights to the planet. Scientists now know that the surface temperature of Venus appears to be as high as 750°K (890°F) and the atmospheric pressure at the surface is nearly 100 times that of the earth's atmosphere. Entirely composed of carbon dioxide and devoid of any water vapor and full oxygen, there is no chance whatsoever of discovering terrestrial types of life.

Today, scientists know very little about the geology and surface characteristics of Venus because of the dense

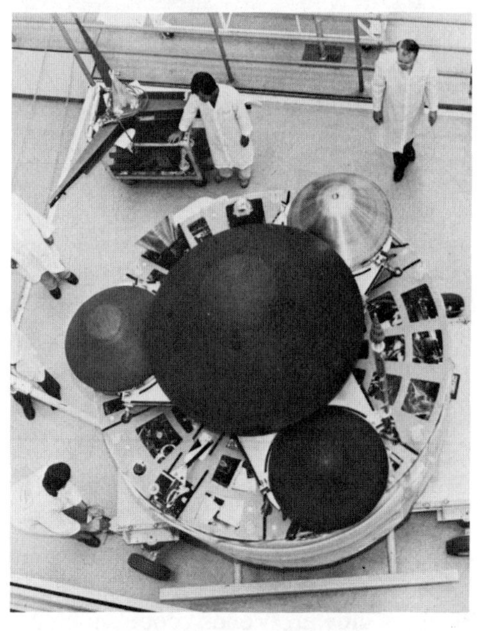

Pioneer-Venus test model.

Pioneer-Venus spacecraft as it appears in flight.

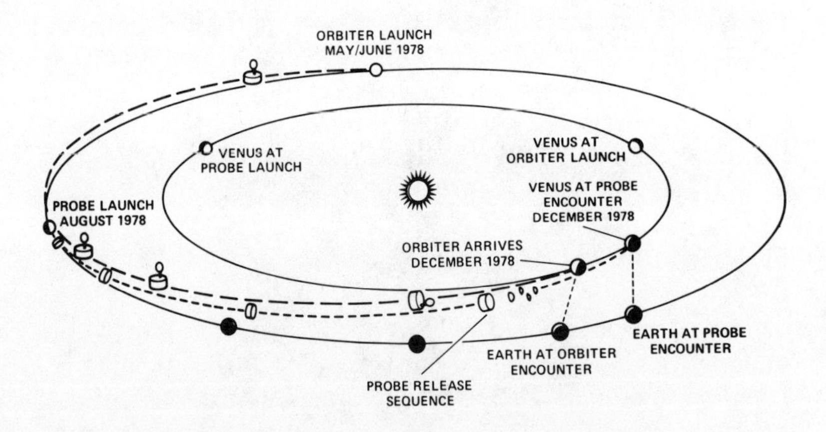

Diagram showing Venus probe sequence.

cloud cover. They, of course, know that the surface is dry, since water must evaporate at the very high temperature existing in the atmosphere. The absence of a strong magnetic field around Venus suggests that its core is considerably different from the earth's core. Recently, there have been some exciting results from an experiment to obtain a topographical map of a 900-square-mile (1500-square kilometer) area near the equator of Venus. Radar ranging was used with deep space tracking antennae. What the map revealed was a giant crater some 100 miles across (160 km) with a rim which rose approximately 1650 feet (500 m) above the surrounding relatively level geography. Such a discovery tells scientists that they may find Venus also pockmarked with ancient impact craters much like those observed on the moon and Mars.

The question astronomers and space scientists hope to answer is why the twin planets Venus and the earth are so different. After all, they were formed around the same time out of similar materials, and they are situated at comparable distances from the sun. Yet, they somehow evolved along very different paths. It may be that the difference is the consequence of Venus having been formed slightly closer to the sun than the earth. At the time of origin, it's possible that the earth's temperature was 280 K (50°F) and Venus's was about 320°K (120°F). Such a difference in the first temperatures could easily have been enough to allow the two planets to follow their own separate styles of evolution.

FUTURE EXPLORATION

To space scientists and officials, the greatest immediate interest for the exploration of Venus rests in better understanding the characteristics and dynamics of its atmosphere—a cloud-filled atmosphere which circulates around the planet every four days, although the planet itself rotates only once every 243 days. Sweeping within 3600 miles of

Venus in 1974, the Mariner 10 spacecraft collected some surprising new information about the dense atmosphere. Photographs taken through ultraviolet filters showed a series of bands and spiral markings roughly parallel to the equator. These markings suggested atmospheric movements in Venus similar to those on earth, except that Venus apparently has no storms.

During the late 1970s, NASA scientists will launch two Pioneer-class spacecrafts to further study Venus. One spacecraft will contain several probes which can be deployed during the planet's approach so that each probe can enter separate regions of the atmosphere. As they penetrate the heavy atmosphere, each probe will measure the cloud composition and observe the structure and circulation patterns which exist. The other spacecraft will orbit Venus to make complementary observations of the atmosphere and ionosphere from space. The orbiting spacecraft will also be equipped with a radar altimeter which can detect gross terrain features. Although the resolution of these observations will probably be poor, the information will enable scientists to develop an accurate, but rough map of the entire planet which illustrates its major surface features.

Scientists feel that Venus holds clues to the earth's weather. When the two Pioneer spacecraft arrive to probe the murky atmosphere, the information on Venus's weather patterns transmitted back to earth may provide clues to the mysteries of the earth's weather systems. On earth, the basic courses of weather patterns are not clearly understood because many factors complicate earth's meteorology. Mixing of oceanic and continental air masses, partial cloud cover, and rapid planet rotation make it hard for scientists to analyze the atmosphere.

Venus, however, is easier to study because it contains a basic atmosphere of 95 percent carbon dioxide and has no oceans and very little tilt to its axis. If space scientists can understand how these variables affect the atmosphere of our closest planetary neighbor, they believe they will be able to define more precisely the impact of the numerous variables in the earth's weather system. NASA hopes that further knowledge of basic weather processes will emerge not only from intensive observations of the earth itself, but also through first-hand studies of Jupiter's fast-spinning atmosphere and Mars' easily observed, largely cloudless atmosphere which is sometimes rendered so opaque by very large dust storms that the surface can no longer be seen.

The Orbiter will be inserted into Venusian orbit in December 1978. The multiprobe spacecraft will be launched in August and the probes will enter the Venusian atmosphere six days after arrival of the Orbiter. The spacecraft consists of a bus, a large probe, and three identical small probes, each carrying a complement of scientific instruments. The probes will be released from the bus twenty days prior to arrival. The large probe will conduct a detailed sounding of the lower atmosphere, obtaining measurements of the clouds, the atmospheric structure, and the atmospheric composition. Primary emphasis will be on seeking data on the planet's energy balance and clouds. Wind speed will also be measured during the descent. The three small probes entering the atmosphere at widely separated points will yield information on the general circulation pattern of the lower atmosphere. Since scientists suspect that the crucial motions are global, only a few observations are necessary. The probe bus will provide data on the Venusian upper atmosphere and ionosphere down to an altitude of about 75 miles (120 kilometers) where it will burn up.

The Orbiter mission is designed to globally map the Venusian atmosphere by remote sensing and radio occultation, and directly measure the upper atmosphere, ionosphere, and the solar wind/ionosphere interaction. Thus, in combination with measurements made at lower altitudes by the large and small probes and the probe bus, plus previous signal measurements, Pioneer Venus will provide a detailed characterization of the planet's entire atmosphere. In addition, the Orbiter will analyze the planetary surface by remote sensing, utilizing radar mapping techniques. This should provide important data on Venus' cratering and surface structure, plus an estimate of global shape. The Orbiter will be placed in a highly inclined elliptical orbit with the lowest point in Venusian mid-latitudes at about 120 miles (200 km) altitude. Operation in orbit should allow investigation over at least one Venusian year (225 earth days). The key questions which the probes hope to answer are: Why have the other terrestrial planets taken different evolutionary paths than the earth? What are the stabilizing and destabilizing feedback mechanisms which determined a planet's climate? Did water ever flow on the surface of Venus? What, then, happened to this water?

NASA scientists feel it would be desirable to repeat the Pioneer Orbiter and probe missions to Venus over a period of years at each mission opportunity. Venus mission opportunities occur at approximately 1½ year intervals. For successive missions the spacecraft would carry instruments designed to answer these specific questions, some of which will undoubtedly have evolved from observations and discoveries made during prior flights. Answers to these questions will expand man's knowledge about the composition, structure, and circulatory patterns of the atmosphere from Venus' outer limits down to near the planet's surface. Later it would be desirable to place a buoyant station within the atmosphere to measure directly and over a long period of time the atmospheric temperature, pressure, and circulation. Such a spacecraft would be designed to remain buoyant like a balloon in the very dense lower atmosphere, drifting on the winds about the planet.

After developing a gross map of the planet's major features using the radar altimeter and the Pioneer Orbiters, the next step would be to more accurately define the surface features with high-precision radar mapping from orbit. The need for a 3-axis stabilization system with high pointing accuracy for the precision radar mapper would dictate a Mariner-class spacecraft for such a mission. Surface details could be observed with about the same degree of resolution as the Mariner photographs of Mars.

Having completed the atmospheric observations from within the atmosphere and the surface mapping from orbit, scientists would then land an instrument capsule on the surface. Such a landing capsule would have to withstand the very dense, high-temperature atmosphere. Such a mission would provide a great deal of information on the physical, chemical, and magnetic properties of the surface; characteristics of the subsurface; seismic activity of the planet; and variations in the static and dynamic conditions of the atmosphere and surface. Information on the data of the Venusian soil could add to the understanding of the history and present condition of the solar system, as well as contribute toward understanding the history of the earth's evolution and how it can be expected to evolve in the future.

JUPITER AND SATURN—THE OUTER PLANETS

Planets of the solar system consist of two types: small, dense, inner planets with solid surfaces such as Mercury, Venus, earth with its moon, and Mars; and, large, mainly gaesous outer planets such as Jupiter, Saturn, Uranus, and Neptune, with some satellites as large as the smaller inner planets.

The first decade of space exploration concentrated on the inner solar system, but at the beginning of the second decade scientists and space technologists started to look at missions to the outer planets. The old fascination of mankind—brilliant Jupiter—became the target for the first mission beyond Mars and Venus.

There is one broad and basic question, dealing with the formation and evolution of the solar system, which exploration of the outer solar system may help answer. Scientists believe that the answer rests in the outer solar system because the inner planets have gone through a great amount of evolution during the previous four billion years. Today, what we observe on their surfaces and in their atmospheres is the end product of long evolutionary processes. Because of the low temperature prevalent in the outer solar system, bodies in the outer solar system have evolved so slowly that their conditions today are not very different from what they were at the time of formation. By exploring that region of the solar system, space scientists are actually going back into time and sampling the characteristics of primitive solar nebula from which the sun and planets are believed to have condensed.

Man's initial steps in the exploration of the outer planets began with the launch of Pioneer 10, which reached Jupiter on a flyby mission in December of 1973. As this spacecraft passed safely through an asteroid belt, it demonstrated that the concentration of dust particles there was no greater than it was closer to earth. The revelation, said NASA scientists, was totally unexpected, and was as puzzling as it was revealing. The flyby also revealed that Jupiter's radiation belts are 10,000 to one million times as intense as those of the earth. Indeed, had Pioneer 10 come 65,000 miles (110,000 km) closer to the planet, the intensity of the radiation would almost centainly have melted the vehicle. The radiation belts appear to be formed similar to a flat disc and held in a magnetic field four million miles in diameter, opposite in polarity to that of the earth's. (A compass on Jupiter would point to the south pole instead of the north.)

Pioneer 10 also yielded considerable data about Jupiter. For example, the planet is an unusual planet by terrestrial standards, both in size and compositon. Only slightly denser than water, Jupiter is 317.8 times more massive than earth. Secondary only to the sun itself, the giant body dominates the solar system. Its gravity affects the orbits of other planets and may have prevented the asteroids from coalescing into a planet. Hundreds of comets are pulled by

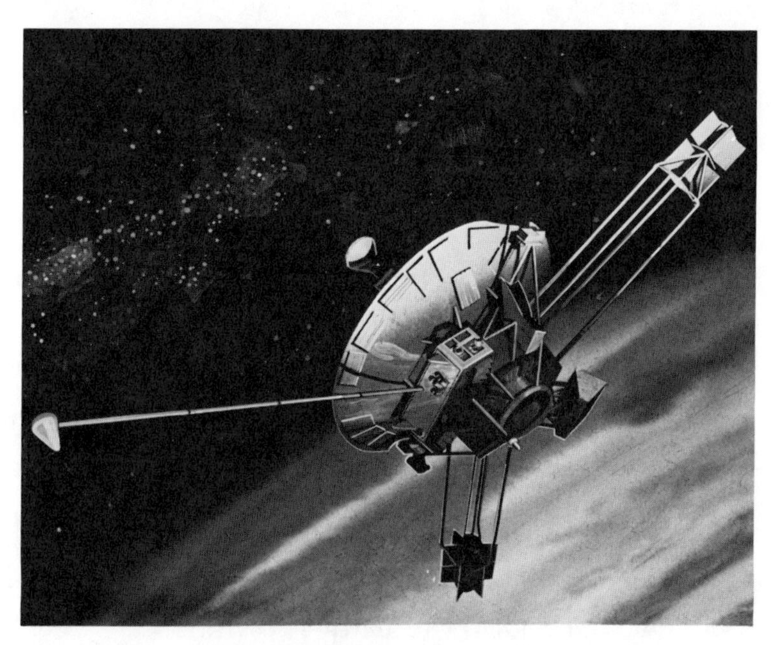

Pioneer-Jupiter spacecraft.

Shuttle Orbiter launching Jupiter probe in space.

Pioneer 10 and 11 events.

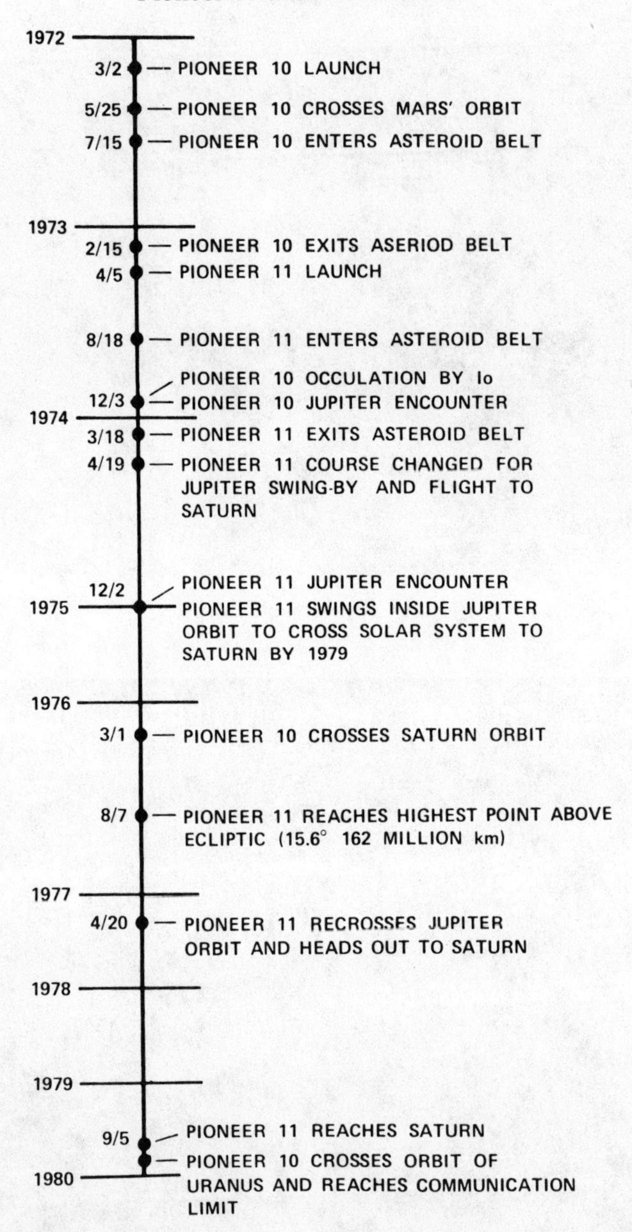

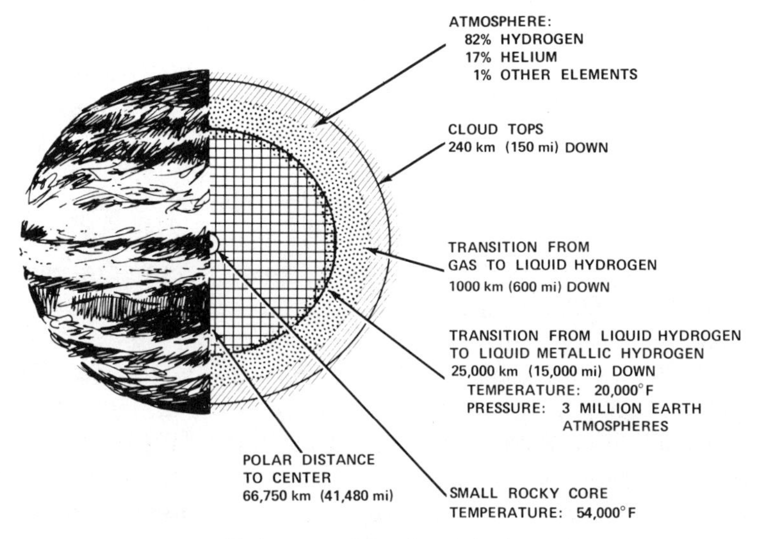

Cutaway of Jupiter interior.

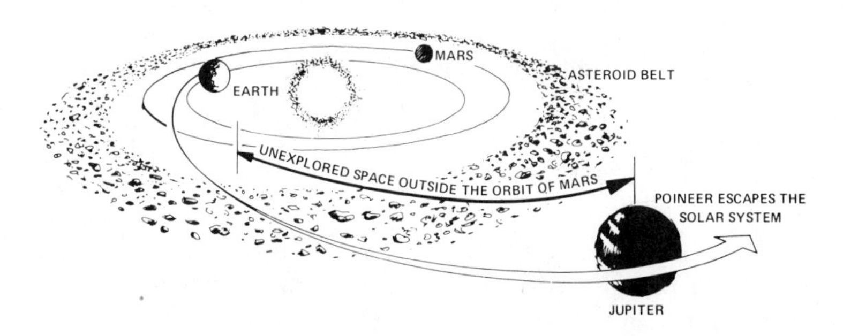

Pioneer route past Jupiter.

Jupiter into distorted orbits, and some of the short period comets appear to have become controlled by Jupiter so that their orbits have their most distant points from the sun about the distance of the orbit of the great planet. Although Jupiter is huge, it is not big enough to have become a second sun, being too small for its own weight to raise its central temperature high enough for a nuclear reaction to be triggered in its core. Scientists say that had Jupiter been 60 to 100 times its present size, the solar system might have become a binary star system, similar to so many other stellar systems. Had this occurred, earth would never have experienced night. Now, Jupiter emits several times more energy than it receives from the sun, energy probably derived from continued cooling of the planet following its primordial gravitational collapse eons ago when the solar system formed. Space scientists feel that a continuing gravitational collapse at a present rate of 1 millimeter per year could provide the observed heat output from Jupiter.

The planet measures 82,967 miles (133,516 km) from pole to pole compared to the earth's 8,000 miles (12,900 km). Rotating faster than any other planet in the system, Jupiter revolves on its axis once in 9 hours and 55½ minutes. This means that any point on Jupiter's equator moves at 22,000 miles per hour (35,400 km) compared with 1,000 miles per hour (1,600 km) for a point on the earth's equator.

Because of such rapid rotation, Jupiter's equatorial regions bulge outward under centripetal force to make the equatorial diameter of the visible globe about 5,767 miles (9,280 km) greater than the polar diameter. Therefore, Jupiter is not really a sphere but has an oblate shape, its polar diameter being 94.2 percent of its equatorial diameter. The earth, on the other hand, is flattened at the poles, but proportionately much less (only 99.66 percent). Although Jupiter's volume is 1317 times more than earth, its mass is only just under 318 times earth's mass. Since Jupiter is much less dense than earth, only 1⅓ times as dense as water, it cannot be a solid sphere like the earth but instead must consist mainly of gas and liquid with possibly a small solid core. At least three-fourths of Jupiter probably consists of the lightest gases, hydrogen and helium (the same gases that are most common in the sun and the stars). Jupiter is thus probably more like the sun in basic composition than the earth.

From a telescope on earth, Jupiter is seen as a magnificent banded disc of turbulent clouds with stripes parallel to the planet's equator. A great mass of dark gray regions sit on each pole in a somewhat transparent hood. The light and dark brown, or gray stripes are called belts. The yellow-white, lighter color bands between the belts are called zones. Although all the colors are muted and soft, they are nonetheless definite. Many of the belts and zones have permanent enough features to be given names. Interestingly, the colors are known to change over the years. The zones appear to vary from yellow to white, while the belts change from gray to reddish brown. The bands fade and darken as well as change color. They also appear to widen or become narrow, as well as fluctuate up and down in latitude—that is, farther from or closer to the equator. Soviet astronomers believe that the cold tops of the clouds in the zones consist of ammonia vapor and crystals. Water clouds probably form at a level too deep in the atmosphere to be identified from earth. A transparent atmosphere seems to rise from 30 to 40 miles (50 to 65 km) above the cloud tops.

Many smaller features suggest unusual details to the zones and bands such as streaks, arches, loops, wisps, patches, lumps, plumes, festoons and spots. Some are probably nothing more than knots of clouds. Such small features often change form rapidly during the course of days, and even minutes. And, because the scale of Jupiter is so vast, the features are thousands of miles in extent. Jupiter's cloud features circle the planet at varying speeds. For instance, a tremendous equatorial current sweeps around the planet at over 250 miles per hour (360 km) faster than regions on either side of it. It represents a 20° wide girdle around the planet.

But of all the fascinating features on Jupiter, the most curious one of all rests in the southern hemisphere. It is a huge, mysteriously long oval termed the "Great Red Spot." At least 15,000 miles long (24,000 km), it has at

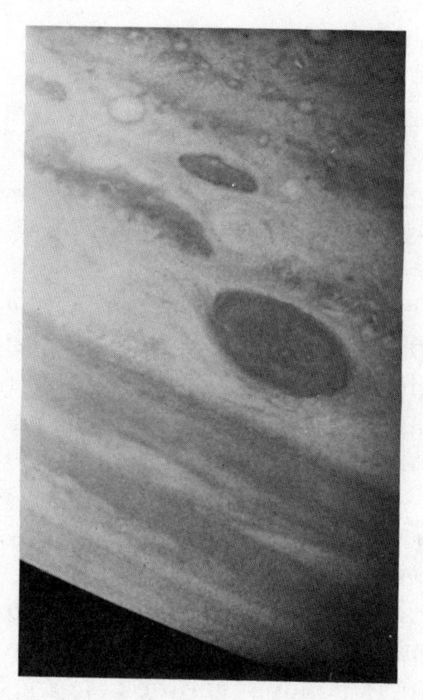

Closest picture taken of Jupiter's "Great Red Spot" by Pioneer 11 from 545,000 kilometers (338,000 miles).

Jupiter from Pioneer 11.

varying times extended itself almost 30,000 miles (48,000 km). The red area has captivated generation upon generation of astronomers since first being observed and recorded centuries ago. Called the "Eye of Jupiter" as early as 1665 by the Italian astronomer Cassini, the spot appeared to vanish from time to time. In 1883, redness appeared to fade totally. Within a few years, it became very distinct again. Early American astronomers felt that it might be a raft or an island, or something floating in the atmosphere, since the spot drifted around the planet relative to the average movement of the clouds. And, often cloud currents swept around it as though the spot itself were a vortex in the atmosphere. More recently, scientists guessed that it really is nothing more than a column of gas, the center of an enormous whirlpool-like mass of gas rising from deep in the planet to the top of the atmosphere and anchored in some way to the surface far below.

Today, however, scientists know the "Great Red Spot" to be a hurricane-like structure, a fantastic grouping of "thunderstorms." Photographs from Pioneer to detect methane show the spot is the highest cloud structure on Jupiter and contains some internal energy source which pushes it above the other cloud layers. This would be unlikely if it were a floating mass such as an island, but could be explained by its consisting of a large grouping of thunderstorms, that is, rising air masses.

White spots have also appeared on Jupiter, although they have tended to be more short lived. They, too, seem to be atmospheric storms because they become bright for relatively short periods of time. The spots also move relative to nearby cloud systems.

As far as the basic internal structure is concerned, the average temperature on the top of the cloud layer is low by earth's standards, probably −190°F. Below the cloud tops the temperature appears to rise steadily. The uppermost regions consist of supercold ammonia crystals, ammonia droplets, and ammonia vapor. As temperature rises with depth into the atmosphere, there may even be ice crystals, water droplets, and water vapor present. Estimates of the total depth of the atmosphere vary enormously from 60 to 3,600 miles before the "surface" is reached. That surface may turn out to be nothing but a gradual transition from gaseous to liquid hydrogen rather than a sharp interface between gas and liquid or a solid surface. Most scientists today suggest a very deep atmosphere at the bottom of which the pressure is enormous, achieving millions of times earth's 14-pounds-per-square-inch sea level pressure.

Scientists feel that such great pressure could convert hydrogen into a very special form in which it behaves like a metal: It readily conducts both heat and electricity as metals do. Thus, speculation is that beneath a sea of liquid hydrogen could be a shell of metallic hydrogen (undoubtedly liquid because of the high temperature) surrounding a small internal core consisting of rocky material and other metals, somewhat the same as the composition of the lunar planets, including earth. Jupiter's core has been estimated as ten times the mass of the earth. However, the existence of such a rocky core is still widely debated among planetologists. Near the center of Jupiter, the temperature might be tens of thousands of degrees and could account for Jupiter radiating into space 2.3 times as much energy as the planet receives from the sun.

Though Jupiter has always been considered so inhospitable that life could not survive, scientists have been revising their thinking. Since there are probably liquid water droplets in an atmosphere of hydrogen, methane, and ammonia, Jupiter may actually provide what scientists term "a primordial soup" in which life can begin, similarly to what happened on earth. Some scientists define life as an unexplained ability to organize inanimate matter into a living system which perceives, reacts to, and evolves to cope with changes to the physical environment that threaten to destroy its organization. As early as 1953, a mixture of hydrogen, methane, ammonia, and water vapor (the kind of atmosphere Jupiter has) was bombarded in a laboratory with electrical discharges. These were passed through the gas mixture to simulate the effects of bolts of lightning. The electrical energy bound together some of the simple gas molecules into more complex molecules of carbon, hydrogen, nitrogen, and oxygen of the type believed to be the building blocks for living systems.

Apparently during some period in earth's early history (perhaps 3.5 billion years ago) something organized the complex carbon-based molecules of earth's oceans and atmosphere into living systems which then managed to recreate or reproduce themselves. From then on, believe the scientists, biological evolution produced all the living organisms on earth, including creatures and man.

But the question which defies answer remains: Has life ever evolved in Jupiter's atmosphere? Today, space scientists know that the temperature may be adequate at lower elevations in the atmosphere. Jovian gas mixtures may be suitable. Electrical discharges are undoubtedly taking place. Thus, Jupiter, perhaps more than any other planet, could contain the answer to the mystery of life. If more unmanned probes are dispatched to that incredibly dense atmosphere later this century, man may solve his most intriguing enigma. Such probes are technologically easy today as a result of the vast experience gained with the Pioneer flybys.

With the launch of two additional Mariner spacecraft for additional closeup observations of Jupiter in late 1977, the next step will be to probe the atmosphere of Saturn employing a Pioneer spacecraft. The mission will be to obtain direct measurements of the structure and circulatory patterns of the atmosphere's upper segments. The Saturn atmosphere has been chosen for such a first probe mission because it has much less density than Jupiter's atmosphere. This means that the design of the probe is within present NASA reentry heating technology. What scientists discover about the atmosphere of Saturn will add a great deal

Saturn as seen through the 120-inch telescope at Lick Observatory, Mt. Hamilton, California.

toward their understanding of Jupiter's atmosphere. Before it is practical to launch a probe into Jupiter's atmosphere, significant breakthroughs in heat resistant materials will be needed.

Once Saturn's atmosphere has been penetrated, scientists will focus their attention on Jupiter once again as the two Mariner spacecrafts place themselves in orbit around the giant. One of the vehicles will be situated in an equatorial orbit with its major goal being to gather imaging data on the Galilean satellites which, hopefully, would reveal their surface characteristics and provide clues to their possible composition and origin. The other spacecraft would be placed in an inclined orbit to map Jupiter's magnetosphere and magnetic field. Such data would lead to an understanding of the interaction between the environment and solar wind of the planet.

Scientists hope by the early 1980s that reentry technology would be sufficiently advanced to allow Pioneer missions to probe the Jupiter atmosphere in even greater depth and clarity. Such missions would provide the initial *in situ* measurements of circulatory patterns, temperature, density, and composition of the massive atmosphere. The missions would then be followed by Mariner spacecraft placed in orbit around Saturn in order to collect the same kinds of information about that planet and its satellites, just as the earlier Mariner Orbiter obtained on Jupiter.

One of the most spectacular objects in the whole solar system awaiting exploration during the late 1980s is Titan, Saturn's largest moon, 2900 miles (4800 km) in diameter and the only moon to have a significant atmosphere. Observations have led scientists to believe that Titan's atmosphere may be ten times more massive than the atmosphere of Mars. Apparently, what is now happening there probably occurred on the earth several billion years ago during the initial steps in life's evolution. It's more than possible that Titan may contain all the insights necessary for understanding the kind of primordial "soup" which formed the solar

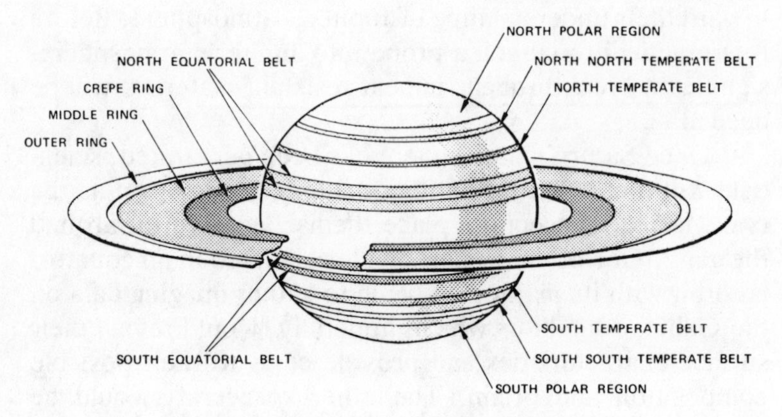

Diagrammatic map of Saturn features.

system, and life on earth. The information learned about Titan and its atmosphere by the outer planet flyby will aid in designing a Viking-type instrumented capsule by 1990 that would be capable of orbiting Titan and landing on its surface.

An option to landing on Titan would be to place the capsule on Ganymede. The objective would be to learn about the composition and temperature of its surface and the nature of its terrain to discover how Ganymede originated and how it relates to Jupiter. Scientists feel that by 1990 NASA will have achieved knowledge about the characteristics of all the planets and their satellites in the solar system.

NEPTUNE AND URANUS

Baffling planetologists as much as Jupiter and Saturn, have been the two most distant giant outer spheres, Neptune and Uranus. For instance, the equatorial plane of Uranus and the orbits of its five satellites are nearly perpendicular to each other, a mechanical aspect little understood. NASA's first opportunity to explore the physics and chemistry of the two planets will occur in late 1979. At that time, Jupiter and Uranus will be so aligned that a spacecraft can be launched to fly past Jupiter and take advantage of its strong gravitational attraction to accelerate sufficiently to swing out past Uranus. A second chance to visit the planet will occur in 1980 when Saturn and Uranus are aligned in such a way that will allow the use of Saturn's strong gravitational attraction to push the vehicle out past Uranus. The missions would revolve around simple observations of the planet and its moons. Since imaging information would be an important part of the mission, a 3-axis stabilization capability for the spacecraft would be necessary. For the second Uranus flyby the goal would be to launch a probe into the atmosphere to measure its composition, temperature, pressure and circulatory characteristics. Thus, the two missions would yield an understanding of the main characteristics of Uranus and its moons as well as answering the puzzling questions about the differences between the Jupiter-Saturn pair of outer planets and the Uranus-Neptune pair. Also, because of the similarity of Uranus and Nepture, the missions would provide a considerable amount of insight about Neptune.

A third opportunity to view the huge distant giants up close will occur in 1986. At that time, Uranus and Neptune will be so aligned that flybys of both planets could be made by one single mission. Since imaging data is of crucial importance for such a mission, a Mariner space vehicle would be used for this planetary visit of almost 3,000,000,000 miles. But without the assist of a Jupiter or Saturn swingby, the total journey using shuttle, space tug, plus a conventional propulsion stage would be approximately ten years. Scientists feel, however, that the information to be gained about Uranus and Neptune would add considerably to an understanding of the origin and evolution of the solar system. Also, the data would enhance knowledge about man's understanding of the sun-planet relationships, which is so crucial to knowing how the sun affects life on earth.

ASTEROIDS, INTERPLANETARY ENVIRONMENT, AND COMETS

For space scientists, complex investigations of the planets are not enough. To them, solar system exploration also includes mapping the interplanetary medium and the solar wind and defining the actual properties of the two other major types of bodies within the system, asteroids and comets.

The interest of scientists in asteroids and comets is due to the belief that they carry the basic ingredients from which the solar system was formed. Comets are believed to have condensed from the solar nebula in the extreme outer reaches of the solar system while asteroids were similarly created in the inner area between Jupiter and Mars. Although primarily created of minerals and rock, asteroids did not melt after they were formed like the larger bodies. This is because their gravitational and radioactive energy was not capable of providing the melting necessary. Thus, asteroids and comets are thought to have remained in their initial form. Analysis of those rock and mineral objects will allow scientists the opportunity to obtain a first-hand look at the material from which the solar system was created.

Also of extreme fascination to space scientists is the interplanetary environment. Evidence recently obtained suggests that energetic particles emitted by the sun create the solar wind which flows outward through the solar system at supersonic velocities and interacts with the planets. This theory suggests that these energetic particles, which form the solar wind, are caught in radiation belts around those planets having the strongest magnetic field. These trapped energetic particles then interact with the ionosphere and upper atmosphere of the planets. Scientists want to develop a reliable concept and model of the solar wind characteristics and flow throughout the solar system, and determine how that solar wind affects other planets and interacts with near-space environments. Knowing this will help them better understand how the sun affects the earth and its climates.

Even though solar wind has been analyzed around the earth for over a decade now, its characteristics near the sun, as well as at the outer edges of the solar system, are obscure. As the spacecraft of the 1980s and 1990s fly past Mars, Venus, and Mercury to the outer planets, each will measure the properties of the interplanetary medium along its flight path. In this new space, the vehicles and instruments will be exposed directly to interstellar wind and different types of magnetic fields. Like the planet earth, the sun is believed to possess a magnetosphere. Scientists know that as the solar wind flows around the earth, it sweeps the earth's magnetic field out into an extended tail. The current theory is that the energy of particles and fields from stellar space is believed to have an analogous effect on the sun's magnetosphere somewhere beyond Jupiter or Saturn. The flights to Uranus or Neptune should aid in the discovery of its location.

Apparently, comets were created as part of the solar system and now orbit the sun. Every once in a while, disturbances provoked by a star, or by one of the major outer planets passing near them, changes a comet's orbit causing it to move deep into the solar system closer to the sun. Such infrequent visitors are the only known comets to the inner solar system.

Although scientists understand very little about comets, they do know that a comet has a nucleus surrounded by a bright coma in the center and a long visible tail created out of gaseous and interstellar dust. A popular belief is that the nucleus is really nothing more than an icy conglomerate of gases such as ammonia and others which have condensed and frozen. That coma is believed to be a gaseous envelope emitted from the nucleus.

Because so little data exist on them, the questions of scientists are basic. For instance, they wonder whether the nucleus is solid or gaseous. What is a comet's composition? How was it formed? Could it have come from another solar system and have been captured by man's sun? To answer such questions means intercepting or rendezvousing with one in order to measure its composition and observe the structure of the coma, nucleus, and tail.

Therefore, a mission of tremendous interest to space scientists will be a rendezvous with Halley's comet when it enters the inner solar system in 1985 on its way to the sun. Halley, with its 76-year period, is a comet which has passed near the sun relatively few times and thus would be rich in primordial substance. It seems that each time a comet passes close to the sun it loses some of its volatiles, which include the primordial matter. The fewer times a comet has passed near the sun the richer it is. Halley's comet is therefore a prime target for investigation.

The asteroids which orbit the sun in a great belt between Mars and Jupiter number in the thousands. Varying from tiny particles to bodies several hundred miles in diameter, the asteroids were too small to have a sufficient amount of radiation to undergo melting when they were formed. Not only is it possible that the asteroids hold in their original forms the materials from which the planets were formed, but they also could contain prebiological organic matter. The great dream, of course, is to obtain a sample of an asteroid. Scientists plan to first fly past a major asteroid, later rendezvous with one, and by 1990 land a spacecraft on one to sample the surface. A separate mission is not necessary to carry out the flyby step since spacecraft on a comet rendezvous mission will pass directly through the asteroid belt. Employing the on-board propulsion and guidance capability, the vehicle will be programmed to pass near one.

THE OUTER PLANETS

Almost nothing is known about the solar system's most distant planet Pluto, which seems to be about the size of the earth and is more dense than the outer planets. But recently planetary scientists have turned and focused their huge telescopes on such outer planets. From their observations, they are beginning to obtain some fascinating data. For instance, as mentioned, Jupiter is a most active planet, emitting four times as much energy as it receives from the sun. This leads scientists to conclude that Jupiter has internal sources of energy which could even be gravitational in nature. Indeed, Jupiter is certainly big enough to be considered a miniature star and is an extremely intense source of radio waves. The energy contained in its bursts is equivalent to a one megaton hydrogen bomb exploding every second. From such incredible measurements, it is theorized that Jupiter is surrounded by an intense radiation belt caused by a magnetic field over 100 times the intensity of that on earth. And, strangely, similar properties have not been detected for Saturn, another outer planet. Scientists wonder if it's because they are too far away to observe these characteristics on Saturn, or if Saturn is totally different from Jupiter?

Earth-based telescope observations have raised more questions than they have answered. Yet, such complex questions enable scientists to plan their exploratory missions to the outer edge of the solar system with the kinds of delicate instrumentation which will provide insights into the puzzling questions about the planets and their satellites.

ALL THE WAY TO THE SUN

For over a decade now, Pioneer 6 has been circling the sun and returning valuable data. That marvel of technological engineering has achieved the longest operating life ever at-

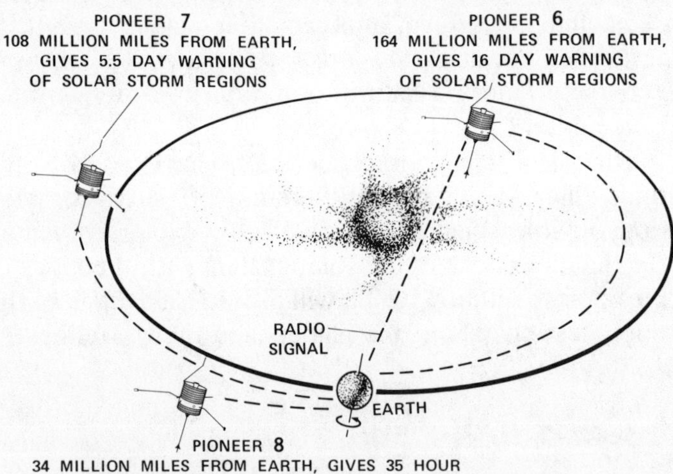
PIONEER 7
108 MILLION MILES FROM EARTH,
GIVES 5.5 DAY WARNING
OF SOLAR STORM REGIONS
PIONEER 6
164 MILLION MILES FROM EARTH,
GIVES 16 DAY WARNING
OF SOLAR STORM REGIONS
RADIO
SIGNAL
EARTH
PIONEER 8
34 MILLION MILES FROM EARTH, GIVES 35 HOUR
WARNING OF SOLAR STORM REGIONS

Diagram of Pioneer 6, 7, and 8 positions on November 20, 1968, as Pioneer 6 passes behind the sun and measures corona.

tained by an interplanetary spacecraft, although its design qualifications called for a required life of only six months.

Pioneer 6 made the first detailed measurements of the interplanetary medium covering 500,000,000 miles. During the years of work, the small vehicle measured the sun's corona, returned information on solar storms from the inaccessible, invisible side of the sun, and measured a comet's tail. Along with these achievements, it has made discoveries about the sun, solar wind and cosmic rays, and the solar magnetic field, all three of which extend beyond the orbit of Jupiter.

Along with its three sister spacecraft, Pioneer 7, 8, and 9 (also years beyond their six-month design lives), Pioneer 6 makes up a network of solar weather stations which circle the sun millions of miles from each other. All of the Pioneers 6 through 11 are still functioning and were designed as rugged, relatively simple, low-cost spacecraft.

Since launch in December 1965, the 140-pound Pioneer has circled the sun almost twelve times, covering over 6 billion miles. During its life, the spin-stabilized, solar-powered craft radioed back measurements twenty-four hours a day from all sides of the sun. Because more recent missions have required tracking time on the "big dish" antennas of NASA's Deep Space Network, Pioneer 6 has generally been pushed aside. Nonetheless, over half of the 50 billion data pieces have been received on earth. That data is then transmitted to Pioneer experimenters, to other space scientists and to the "solar weather forecasters" of the National Oceanographic and Atmospheric Administration's Solar Disturbance Forecast Center at Boulder, Colorado. There, measurements by the four Pioneers whose positions on various sides of the sun change constantly are used to forecast solar storms for some 1,000 primary users. These include the Federal Aviation Agency, commercial airlines, power companies, radio and television communication companies, military organizations, and groups working on surveying, navigation, and electronic prospecting.

Drum-shaped, some 35 inches high and 37 inches in diameter, Pioneer 6 is covered with solar cells and is divided by a narrow circular band with openings from which four experiments and four orientation and timing sun sensors can scan during rotation. Three booms 120° apart provide its own data-handling temperature control, communications, and power system. All told, the little vehicle has some 56,000 parts.

A key role played by Pioneer 6 was the charting of the solar wind and the twisting magnetic fields threading it, plus the twisted streams of high energy particles which the magnetic field pulls out from the sun. Such phenomena corotate with the sun every 27 days. Masses of such data have allowed better understanding of the solar corona from which the solar wind "boils off" into interplanetary space and of the sun itself. Pioneer also measures little pieces of the sun, that is, the particles thrown off into space. Measurements have provided new understanding of the earth's magnetosphere which shields earthlings from high energy particle radiation. As the tiny spacecraft passes behind the sun, it emits radio signals which pass through the solar corona. This adds to knowledge of the corona.

All four sun-orbiting Pioneers can help predict solar storms when they are positioned behind the sun, since they can observe events on the solar surface up to three weeks before they become visible from earth due to the sun's 27-day rotation on its axis. Scientists feel that such "geomagnetic storms" may trigger earth's long-term weather. They are created by the huge bursts of solar wind which buffet and vastly distort the earth's magnetic field for as long as a week. This, of course, throws circuit breakers, causing power blackouts. It makes compass navigation and surveying impossible and cuts off radio communications overseas.

The incredible longevity of Pioneer 6 allowed time for it to get behind the sun and observe the corona. It also provided for important analyses of the same solar wind stream by two spacecraft, a million miles "down solar wind" from each other. In November of 1969, Pioneer 6, inside the earth's orbit, and Pioneer 7, outside it, were aligned on the same solar wind streams and magnetic field lines. In September of 1970, a similar line up occurred with Pioneer 8. In August of 1972, Pioneer 6 measured effects of the solar shock wave on its side on the sun when three huge solar flares were superimposed on each other during a cataclysmic solar storm. In December 1973, Pioneer 6 studied the tail of the comet Kohoutek as it passed through the solar system. In March 1974, the vehicle measured solar phenomena in a classic three-spacecraft lineup. Pioneer 10 was inside the earth's orbit while the much younger Jupiter and Saturn Pioneers (Pioneer 10 and 11) measured the solar wind stream about ten or eleven days later near Jupiter some half billion miles away. In late 1974 and during 1975, three more measurements were taken. Finally, in September of 1976, because it had lived so long, Pioneer teamed up with the U.S.-German Helios satellite, which was 6,000,000 miles inside the orbit of Mercury, closer to the sun than any spacecraft has yet travelled.

THE NEW TECHNOLOGY NEEDED

In future programs of solar system exploration, major new support systems need to be developed. For example, a Mars surface sample return mission requires the creation of more major new systems and operating capabilities to support them than any other future solar system missions currently being considered. Most important, a new chemical propulsion stage will be required to return the sample to earth. But to many scientists there exists the question as to whether it is really necessary to conduct an assay of the Mars surface sample before it is landed on earth in order to determine the hazards of back contamination. One method by which this might be accomplished is to rendezvous with the sample return module in each orbit and use the Spacelab to evaluate the sample. This on-orbit assay and the subsequent landing of the sample would require three major support systems, the spacetug, shuttle, and Spacelab. The spacetug would rendezvous with the Mars sample module, and return it to the shuttle. Such a rendezvous is likely to occur in a highly elliptical earth orbit to minimize the propulsion requirements of the sample return module. Once the surface sample has been returned to the shuttle, the assay could be handled in a properly equipped spacelab before the shuttle lands carrying the sample. As part of the mission planning, it would be necessary to develop techniques for properly handling the surface sample and for making the on-orbit assay. Of crucial significance is the avoidance of any possible back contamination, while at the same time doing little to destroy the biological value of the sample.

As far as the exploration of Venus is concerned, the development of two important systems is necessary to carry out the Venus Radar Surface Mapper mission. One is a new radar system which will provide high resolution images of the surface that can be adapted to a Mariner spacecraft. The other is a closed-loop altitude stabilization system with high pointing accuracy. It is believed that a pointing accuracy as high as +0.25 degrees is necessary. Closed-loop stabilization system technology developed for the large space telescope would be applicable to the design of such a system.

For the complex Venus Lander mission, the instruments, power system, and communications system must be designed to function in the high temperature and high pressure environment which exists on the surface of Venus. Although the landing module design may be such as to safeguard some of the systems from the nearly 100-atmosphere-pressure environment, some instruments must be necessarily exposed to it. The high temperature (890°F) on the surface will be the most difficult problem with which the designers must deal.

Probing the atmospheres of Jupiter and Saturn requires creation of techniques for measuring temperatures, pressures, and composition of heavy atmospheres when entering them at 50,000 feet per second. The Jupiter probe mission will require a major advance in entry heating technology. The combination of the very dense atmosphere of the planet, which at its depth is more than 1000 earth atmospheres, and the very high velocity of an entry probe will result in entry heating rates far beyond those with which current technology can cope.

Missions to Uranus and Neptune will require almost ten years. Therefore, development of technology for providing ten- to twenty-year long-range duration flights is necessary.

As far as a çomet intercept mission is concerned, a unique system is needed which will be able to collect samples of the comet while traveling through the tail, or the coma at a relative velocity of two to four kilometers per second in order to deliver these samples to the detecting instrument without destroying their natural gas by ionization.

In addition to advanced solar electric propulsion systems for slow flybys and rendezvous with comets or asteroids, a new on-board navigation system is needed. The comet rendezvous spacecraft must be launched before the comet is sighted on its path in toward the sun. Since the exact time and precise location of the comet's return cannot be accurately predicted, the spacecraft requires an onboard navigation şystem capable of determining the velocity correction needed to carry out rendezvous and providing the guidance signals which will enable the propulsion system to make the needed velocity changes.

EXPLORING THE UNIVERSE

Extraterrestrial space activities involving the universe revolve around three timeless riddles which men from all cultures, past and present, have tried to explain: How did the universe begin and develop? How did life originate and evolve? What is man's place and destiny in the universe?

As scientists observe the stars around them, they find themselves faced with an entire panorama of nature and evolution within the universe. To them, scientific problems raised by such questions are as diverse as the origins of man: from inanimate material of the earth's surface, the motion of the continents, the Caribbean's spiralling hurricanes, the creation of the planets, the synthesis of elements in the centers of stars, and the beginnings of the universe itself. Today, all are being pursued employing the same basic laws of physics, chemistry, and biology, and, of course, all are related to the central question of the evolution of the cosmos.

To attain such an understanding requires collecting and analyzing as much of the data as possible that the objects throughout the universe are transmitting to man. The tool most employed to gather such information has traditionally been the telescope, which serves as a collector of the energy or particles radiated from the various objects in the universe. Such radiation includes the entire electromagnetic spectrum from the very lengthy radio waves to the infinitesimally short gamma-rays. Thus, a whole new series of telescopes is required, each with its own collector and detecting instruments tailored to a band of wavelengths.

Only a very small amount of the radiated energy and particles which touch the top of the earth's atmosphere get through the surface to be collected with earth-rooted telescopes. If it does, the resolution is often distorted by the atmosphere and from radiation generated on earth. The solution is to place telescopes in space to analyze the widely varied phenomena of the universe. Even then, most of the data the astronomer collects cannot be explained. For instance, recently discovered pulsars, quasars, and black holes in space are without explanation by even the most imaginative scientists. One of the main reasons man is unable to understand such happenings is due to an absence of an adequately precise theory of gravitation. Only in space can the critical experiments necessary to confirm a more precise theory take place. To scientists, the use of space as a laboratory is the most important source for conducting experiments which will aid in explaining these and other physical phenomena of the universe.

Of all the sciences, astronomy provides the scientists with the fundamental laws of celestial mechanics and helps in the development of mathematical concepts of integration, time, and computers. Most crucial of all, it allows for the developments of the basic concepts of nuclear energy. Who can say what discoveries of major scientific and practical significance await mankind in the future?

Stated simply, NASA's goals, objectives, and possible future achievements in exploring the universe include:

- To utilize the understanding of the universe and the phenomena occurring in it for the benefit of all earth's people.
- To increase knowledge of the fundamental principles and laws of physics which govern observed cosmic phenomena.
- To understand the origin and continuing evolution of the cosmic environment, by observing and interpreting the basic physical processes in the solar system, stars, galaxies, and universe.

Any investigation into the nature of the universe must be based upon a theory of cosmic evolution. Of course, such a theory does not mean that it has been proven, yet it serves as a helpful, simple framework within which earthlings can define general themes of extraterrestrial investigation and to specify objectives to help steer the planning of specific programs and engineering of necessary hardware.

In terms of the basic cosmic evolution concept that most scientists adhere to, the universe began with a gigantic explosion some 15 billion years ago, known as the "Big

Bang." The matter of the universe in the form of the most fundamental particles such as electrons, protons, and neutrons were flung apart with tremendous speed. Apparently within the first few moments of the explosion, the protons and neutrons fused together to create hydrogen, denterium, and helium. As the universe continued its expansion, the matter slowly cooled and condensed into galaxies. Within the galaxies, the matter turned into stars. In the interior of the stars, somehow the temperature was raised again to millions of degrees which allowed hydrogen to fuse in a series of reactions, forming helium first and then many of the remaining substances of the universe. All of the other elements were formed when the more massive stars exploded just before they flickered out. Now, 15 billion years later, the universe seems to still be expanding. Many of the stars which formed at the beginning have since died, possibly to wind up either as infinitely compressed "black holes" or in fantastic explosions which spread the freshly "cooked" material all over the galaxy. From such stuff, new stars were created and evolution took place. One of the 100 billion galaxies is our own Milky Way, which in itself contains 100 billion stars. One of these stars, the sun for earth, was born some 4½ to 5 billion years ago. As the sun developed from a cloud of gas and dust, some material was trapped in orbit around it and gave birth to the planets, including planet earth.

During the millions of years which followed, no living form existed on our simple, frail planet. The atmosphere was probably filled with mixtures of carbon dioxide, nitrogen, ammonia, and water vapor. Either solar ultra-violet radiation or flashes of lightning (or perhaps both) synthesized large amounts of amino acids, sugars and fatty acids and eventually these crucial molecules settled in the planet's oceans. Somehow, unexplained collisions occurred between them, linking small molecules into larger ones called nucleotides. Over hundreds of millions of years, the concentration of these molecules increased so much that the oceans became like a "soup." Out of such substance, a living cell appeared and the first threshold was crossed from the non-living to the living. Most scientists believe that the process could be a common occurrence on other planets in other galaxies, in other Milky Ways, in other universes.

Of course, no one can say for certain what aspects of the "Big Bang" theory are, in fact, true. Although there may be billions of other planets in man's galaxy, scientists do not possess unambiguous evidence of even one planet outside of the solar system. Indeed, life may have originated on many of these planets, but there is only evidence of life on planet earth. Intelligent life may be widespread in the universe, but man has not yet touched it. Many questions, puzzles and uncertainties remain. However, these unifying concepts, in which the expansion of the universe, the birth and death of galaxies and stars, formation of planets, the origins of life, and the ascent of human beings are all explained by the different features of the process of cosmic evolution present an impressive challenge to astronomers, physicists, geologists, and biologists. The most awesome task awaiting them during the next several decades is to test and evaluate this concept of cosmic evolution.

It is entirely possible that while doing this, entirely new theories and concepts will evolve. A new physics, or even a new astronomy and biology may be necessary to explain the various aspects of the universe and the existence of life elsewhere. Today, because of the historical correlation between the development of technique and the development of theory, space scientists are at last able to clarify a number of extremely intriguing questions, critical tests, and even diagnostic measurements for various aspects of the cosmic evolution theory.

Included in these questions are: If such a concept of origin is accurate, where and when did the "Big Bang" take place? What actually happened during the explosion? Did all galaxies form at the beginning, or do they condense, evolve, and die as the stars do? If the "Big Bang" theory is not correct, can evidence be discovered to support an alternative explanation? What exactly are quasars? Is there enough mass in the universe to block its further expansion? What is the nature of gravity? Do black holes really exist? How is the observed energy in quasars being generated? Are physical laws not universal? Do other stars have similar planetary systems? What processes occurred during the creation of the solar system? What drives solar activity? How does it affect earth's climate? What is the prognosis for the sun, and the earth? What actually happened on earth and on the moon during the first billion years? Precisely how did life arise on the earth? Does life, intelligent life, exist elsewhere? Will we ever be able to communicate with this life?

These are extremely simple, yet broad, questions of exceptional scientific interest. The current NASA program in response to these questions is focusing primarily around the current and projected tools of astronomy. Because the light and energy spectrum over which the astronomer can view the heavens from space extends from radio signals, which have a very long wavelength, to the infinitesimally short wavelength x-ray and gamma-ray signals, a whole new family of telescopes, each with characteristics different from the others, is needed to cover the entire spectrum. For instance, the length of radio waves is measured from centimeters to kilometers, while x-rays are in the range of one one-hundred millionth of a centimeter and gamma rays can be ten-thousand times smaller than this. Currently, scientists employ several astronomical tools for the collection of data:

- OPTICAL TELESCOPE Used in ground-based observatories. When used in space, it can observe the ultraviolet rays which are filtered out by the earth's atmosphere. Such an optical telescope requires a large mirror to collect the visible and ultraviolet rays. To obtain the high resolution needed, the mirror must be accurately polished. Light rays collected by the mirror pass through an optical system which concentrates them in a beam that is focused on the light-detecting instruments used to sense the image of the objects being observed.

- HIGH ENERGY TELESCOPES Studies in the high energy spectrum need detectors having a large collecting area which is sensitive to x-rays and gamma-rays. Since high energy radiation is unable to get through the earth's atmosphere, studies in this area of the spectrum must be made from earth orbit. Similar to the optical telescope, this high energy telescope is operated from an earth-based control center. With this telescope, astronomers will be able to develop an understanding of physical processes taking place in the universe, the most important being the changes in the matter and the states of stars.
- SOLAR TELESCOPES With such a telescope, the astronomer can study in detail the structure of the solar surface and the major disturbances taking place on it with the idea of developing an understanding of the processes and mechanisms at work on the sun.
- INFRARED TELESCOPES This telescope opens the door for the astronomer to better understand the cold bodies and dust clouds found in the mass between the stars and galaxies, the nature and behavior of the material and the fields between the stars and galaxies, and the interrelationship between the stars and the interstellar clouds.
- RADIO TELESCOPES A radio telescope having a dimension of ten kilometers and a dipole antenna deployed in orbit allows astronomers to map and analyze the origin of radio waves generated by remote objects in the universe and to compare them with those radiated by the major planets within the solar system. The wavelengths which can be effectively observed with pointed radio telescopes is limited by structural size considerations of the telescope

itself to about 100 meters. The large radio telescope allowing observations up to kilometer wavelengths with good resolution will open the complete radio spectrum.

- SPACELAB Employing Spacelab for astronomy and physics observations during 30-day missions in space is the dream of scientists because observations and experiments made during such missions can support and augment studies made by the various telescopes.

As exploration of the universe begins to accelerate, new tools of astronomy and future mission opportunities are being conceived. For instance, a major advance is the Large Space Telescope (LST) which will allow for optical astronomy observations from space. Types of the LST being considered for possible adoption in 1982 include a three-meter diameter telescope mirror that weighs about 22,000 lbs. The LST will be placed into orbit by shuttle.

Housing a large mirror, the LST can observe further out into space than ever before. Its resolution will allow study of the structure and evolution of remote galaxies, the fundamental characteristics of interstellar and intergalactic matter and fields, as well as the physical nature of stars. From such observations, scientists can learn a great deal about the large-scale structure of the universe in terms of homogeneity, size, curvature, and evolution.

The following tables illustrate a summary of all the opportunities NASA has to investigate the universe, as well as the major technology and system report requirements needed.

Future space activities.

Benefits	Space Disciplines	1976–1980 Automated Satellites	1980–1987 Sortie Missions Manned	1987–1995 Free Flying Space Labs Automated / Free Flying Space Labs Manned	1995–2000 Space Station (Manned
Social & Economical	1. Weather (atmospherics) Solar physics, Ionosphere, Mesosphere, Stratosphere, Tropopause } Interactions energy transport	Obtain raw data that will be used to develop theoretical weather models (sounding; winds; humidity; temperatures)	1. Investigate new instruments and sensing techniques 2. Measure specific parameters in specific locations to aid in the development of theoretical models 3. Basic research in new areas or applications 4. Verification of preliminary theoretical models	1. Extend duration observations (30-60 days) with specific instruments and sensing techniques 2. Theoretical models developed and verified 3. Hardware and operational concepts developed and verified 4. Basic research in new application (Private industry participation)	Operational observatories - Weather modification - Climate modification (Manned & automated)
	2. Oceanography Atmospheric interaction, Ocean currents, Upwelling divergences, Convergences, Ocean dynamics, Ice coverage & dynamics } Energy transport	Raw data for theoretical ocean basin models (temperature; salinity, wind; waves; seastate; currents)			Operational observatories - Weather modification - Coastal region wave forecasts - Ocean states (Manned & automated)
	3. Resource utilization Agriculture - food production Forestry fiber production Crop forecasts Mining & Oil Energy-basic materials Environment Transportation corridors Urban planning Earthquake warning Flood/hydrographic surveillance	Raw data for theoretical resource management models (multi-spectral sensing imaging change detection)			Operational observatories - Resources managed by industry and governments (Private industry participation) (Manned & automated)

Social Economical	4. Communications and navigation Aircraft & ship traffic Rescue Communications	1st generation of comm & nav satellites (Private industry participation)	1. Develop & test new concepts and hardware | (Private industry participation)	1. 2nd generation of comm & nav satellites 2. Development of new hardware & operational concepts (Private industry participation)	- Wide band communications - Interplanetary flight control - Traffic control (Private industry participation) (Manned & automated)
Economical	5. Materials and Space manufacturing Molecular biology Pharmaceutia Exotic materials	This discipline will probably wait for sortie missions	1. Experimental labs 2. Evaluate feasibility	1. Extended duration laboratory facilities (30-60 days) 2. Develop concepts for space factories (Private industry participation)	Operational space factories (Private industry participation) (Manned)
Social Scientific	6. Life Sciences Medical research Biological research Life processes	Collection of raw data Development of hypothesis	1. Experimental labs 2. Basic short term research	1. Extended duration of research facilities	Operational laboratories - Medical & surgical - Biological (Manned)
Scientific	7. Pure Science Astronomy Space physics Planetary exploration	1st generation of satellites - OAO, head, OGO, OSO, explorer, Viking, etc.	1. Develop & test new instrumentation 2. Basic short term research	1. Extended duration research facilities 2. Development of hardware and operational observatories	- Operational observatories - Relay station for exploration (Manned & automated)
	Sensors	1st generation (automatic)	2nd generation (new) sensor development (manned)	3rd generation (new) sensor development (automatic & manned)	Operational sensors (automatic & manned)

VI

Developing the Space Frontier

If I have seen farther than Descartes, it is by standing on the shoulders of giants.

Sir Isaac Newton

Throughout the past two decades, imaginative ideas by creative scientists have become the deadly serious engineering design studies by aerospace companies and their technologists. Suddenly and surprisingly, new hardware to achieve incredible projects in space is available. The settlement of the moon, the construction of huge colonies in orbit, the humanization of space, the recovery and processing of lunar, planetary, and asteroid material, the sending of whole explorer teams to the outer edges of the solar system and the universe are no longer the "might be's" of the future, but the reality of today. Some dreams, although sound in principle, are still many decades beyond current capabilities—such as high-speed probes to other planets, or making other planets habitable. However, with our present technology, we can establish a major lunar base, or an orbiting factory housing thousands of engineers and technicians building such structures as satellite solar power stations. During the years between now and 2000, earthlings will be busy at work developing the space frontier. The projects receiving the most intense efforts include the following.

A LUNAR BASE

A base on the moon is considered the major priority for the humanization of space in the vicinity of earth. Because of the moon's rich and abundant mineral resources, the development of a space station or space colony would achieve economic justification since the resources necessary for such construction would be too costly if brought from the earth.

In its initial form, a lunar base would be built totally from material carried to it from earth. Unprocessed lunar material would be used as the protecting layer, or cover, over the lunar living quarters. The inhabitants would depend upon supply lines from earth. Gradually, with the recycling of air and water, the base would become self-supporting. Lunar rock and soil would provide the materials for construction in space, the substance for crops, fields, and farms, and fuel for nuclear power reactors. With the expansion of lunar base activities and utility needs, a nuclear power station would be established to utilize lunar thorium for conversion to U-233 in breeder reactors. After such steps have been taken to provide for base needs, power would be exported to earth by microwaves or lasers. Soon, multi-gigantic power levels would be achieved.

A variety of activities will take place. Not only would the exploration of the moon continue, but its unique environment would be employed as an excellent site for major radio astronomy facilities, shielded from the radio noise produced by humans on earth. The structure housing the radio astronomy equipment would be very large, since the gravity is low and there is no wind. Not only this, but the moon's slow rotation rate offers advantages for both radio and optical astronomy. Optical astronomy would have advantages similar to those in free space—those that are due to the lack of atmosphere. A lunar science base might also be useful in x-ray and cosmic ray observations and serve as a site for detecting gravity waves.

Space scientists argue that by 1985 many commercially desirable advances will have already taken place. For example, in terms of the manufacture of satellite power stations, they feel that through technological advances the installed cost of a power station in space will be reduced to that of its earth-based competitors—$500 to $1000 per 1978 kilowatt dollar. And, furthermore, a significant part, indeed perhaps all, of the annual increase in installed capacity will be represented by space systems by 2000. In order to appreciate such a concept, the magnitude of man's energy needs should be reviewed. For example, if the demand for electricity grows as it has for the last fifty years, the need for new installed capacity by 2000 will be almost 250 million kilowatts in that single year alone. At $500 per kw, this suggests an annual capital expenditure of over $200,000,000,000. Obviously, a lunar industry capable of manufacturing space power stations would be worth billions of dollars a year.

FUTURE PROGRAM PACE

The pace at which scientists develop space exploration capabilities will be dependent upon two factors: the rate they advance their technological capability to develop and effectively utilize advanced space systems; and the establishment of genuine, nationally recognized needs for such systems. Such priorities can be broken into several groups including needs for services, needs for scientific and technical knowledge, and needs to improve economy of space operations. The following summarizes NASA's future space operating capabilities:

Explorer

- A relatively simple spacecraft which is generally used for special purposes and thus contains relatively few instruments and experiments.
- Spacecraft costs range from a few million up to about $25 million each.

Pioneer

- A spacecraft with more capability than the Explorer in terms of onboard power, data handling capability, and diversity of experiments; this spacecraft is sometimes spin stabilized.
- Spacecraft costs generally range from about $25 to $75 million each.

Mariner

- A relatively complex spacecraft generally having 3-axis attitude stabilization, onboard propulsion, high power and data handling capabilities, and thus the ability to handle a significant number of instruments and experiments.
- Spacecraft costs generally range from $75 to $125 million each.

Observatory

- Large permanent spacecraft able to handle a large number of complex instruments and with the capability to 1) be serviced and refurbished in earth orbit, or 2) carry out observatory type operations in orbit or on the surface of a planet.
- Spacecraft costs range upward from about $125 million each.

MINING MARS AND THE ASTEROIDS

With low-cost, large-lift transportation, costs for placing mining engineers on Mars and the asteroids will be significantly reduced. With nuclear propulsion units, the trip time to Mars and the asteroid belts will fall within convenient, economic periods of travel. Flights of this type are expected to take place routinely as soon as scientists learn how to remain in space for the durations of such flights. But NASA feels there is no question that man will be mining Mars and the asteroids for their mineral wealth by 1990.

For example, an iron-nickel asteroid one kilometer in diameter represents four billion tons of high-quality steel with a value of approximately $425,000,000,000 on the 1978 market. The way in which a four-billion-ton payload from the asteroid belt would be brought into earth orbit includes the use of a high-thrust, high-specific-impulse propulsion device using nuclear energy. The asteroid would then be sawed into small hunks which would be brought down safely through the atmosphere to a special dock. Although mining asteroids seems like something out of Buck Rogers and the 21st century, most scientists feel the concept is within the realm of possibility.

SETTLEMENT OF THE OTHER PLANETS

The very first problem man must solve in settling Mars, Venus, and other planets, revolves around the creation of a breathable atmosphere. The problem is different in the case of each planet. For example, the moon has no atmosphere; Mars is enveloped in an extremely thin atmosphere of carbon dioxide; and, Venus has an extremely thick, or dense atmosphere consisting of carbon dioxide and a hot surface of 750° K.

The two methods for modifying their atmospheres are

(1) biological processes which would convert carbon dioxide into oxygen and some other carbon-containing molecules, and (2) the direct application of energy by nuclear devices. On Mars, the explosion of a small number of nuclear devices could vaporize enough of the perennial northern cap to release enough carbon dioxide into the thin atmosphere to forever establish the greenhouse effect. From that point on, solar energy would evaporate the rest of the Martian cap, producing a substantial rise in the density of the carbon dioxide atmosphere. That greenhouse effect would increase surface temperatures for the growth of plants. If nitrogen and water were introduced into the Martian atmosphere, then plants could slowly create a breathable oxygen atmosphere. Unfortunately, because the process is limited by available sunlight, it would require hundreds of years to accomplish. Perhaps, by 1990, a third modification will be conceived which would create the necessary greenhouse effect more rapidly.

As far as Venus is concerned, huge amounts of carbon dioxide would have to be removed from the atmosphere in order to reduce the greenhouse effect enough to allow the surface to cool. A few scientists have felt that seeding the atmosphere with a form of blue-green algae would change the atmosphere dramatically. But, most agree the atmosphere of Venus would be too hot and deadly for the organisms to endure.

On the moon, oxygen will be produced by heating lunar surface material with nuclear or solar thermal collection units to trigger the release of oxygen from the minerals. Realistically, however, the amount of energy needed to produce a breathable atmosphere in such a manner would be excessive. However, because the present vacuum condition could prove extremely valuable for future lunar operations, there may be real advantages in keeping the moon's atmosphere exactly as it is now.

INTERSTELLAR FLIGHT

Today, a spacecraft launched by current propulsion technology requires at least seven thousand years to reach the nearest star. There is little point in sending one off now because scientists feel that by 1985 they will have developed nuclear rockets capable of cutting the flight time of an explorer to less than seventy-five years.

Some are even speculating on the possibility of creating and harnessing antimatter.* Such a happening represents the most efficient energy storage system yet devised for the total conversion of mass to energy. If such systems could actually be operated as rocket engines, flight time to a star would be cut to fifteen years. Although it is possible to create antimatter in high-energy accelerators, no one has figured out how to store it in such a manner that it could be used as an energy source for a rocket engine. If a means of storing antimatter were to be developed in the next few decades, it would be possible to construct an antimatter reaction rocket motor by 1995. But this would be expensive in terms of energy. The minimum energy needed to create antimatter is equal to the energy released in a matter/antimatter annihilation. NASA officials warn that the manufacture of antimatter particles is in fact a highly inefficient process, the total energy required being much higher than the energy eventually released. And, of course, energy will be costly for decades to come.

FUTURE PROGRAM OBJECTIVES

Other potential space objectives have been defined for the decades beyond 1980 by NASA officials and scientists. The following summary is by no means complete. Instead, it

*Matter composed of the opposites of the elements which ordinarily make up matter—antiprotons in place of protons, positrons in place of electrons, and antineutrons in place of neutrons.

suggests in an abbreviated form how much vision space scientists are demonstrating in their current planning. Their objectives for a universe which not only contains a great variety of things in it, but is also continuing to expand, include the following:

Agriculture

- An operational global crop production forecasting system based upon satellite surveys.
- Space systems required for global crop production forecasting can provide data to support the assessment of arable land; monitor the "health" of the land, rangeland, and forage status; control of insects and disease; timber inventory; and various aspects of forest management. These activities all contribute to food and forest production and management. Specialized analytical techniques are needed, and appropriate effort will be focused on this.
- An operational, global, climate-predicting system based upon satellite surveys. Long-term climatic changes have a major impact on food production. The monitoring of long-term trends could offer a tool to ward off overnight catastrophe or long-term economic regression.
- A system of accurately surveying global estimates of crop yield for improved food management decisions.
- Space techniques and systems to measure soil moisture and soil temperature.
- Water availability forecasting models based upon satellite measurements of snow moisture content.
- An operational marine resources system to measure marine yield and possible extensive cultivation of the marine environment after 1980. Satellites with the ability to survey large ocean areas in short times can assess the environmental/biological relationships upon which a future systems will be based.
- Communication satellites which will carry educa-

tional information to rural areas in order to improve farm practices.

Climate

- Creation of a number of space observation systems which will yield important information on the climate's radiation budget, surface radiation, snow and sea ice, polar ice sheets, hydrological parameters, atmospheric cases and particles, and ocean dynamics.
- Space monitoring systems for the stratosphere.
- Monitoring from space of water quality with measurements relayed by satellite to analysis centers.
- Satellites to predict local and severe weather patterns.
- Measurement systems for tropospheric pollutants.
- Space sensing systems to provide environmental information leading to the control and eradication of harmful insects.
- Hazard forecasting systems for forest fires, floods, and similar disasters.
- Earthquake prediction satellite systems measuring dilatancy and plate motions.

Energy and mineral explorations

- Space systems to aid mineral exploration.
- Space solar power stations to contribute to man's energy needs.
- Satellites for earth point-to-point relays of power.
- Nuclear waste disposal outside the solar system.
- Earth monitoring systems for help in determining sites for new power plants, strip mines, and the long term effects of pollutants on the atmosphere.

- Satellites for mineral exploration which can help to develop geological maps, reference magnetic surveys, and mobile position determination.

Consumer services

- Establishment of forty intercontinental and domestic communication satellites by 1985.
- Communication satellites meeting the education and health care needs of remote and rural peoples.
- Direct TV broadcasting satellites.
- Household alarms of impending disaster, storms, etc., triggered by special satellites.
- Satellite-based navigation systems for larger ships and aircraft; such high-powered satellites would allow extension to smaller boats and other mobile devices economically.

Scientific commercial services

- Skylab will yield material property limits to stimulate progress in ground-based techniques, production of materials with improved properties for ground processing, new materials with unique properties, and the potential for commercial processing in space.
- The study of the zero-gravity time course of disease in gravity sensitive systems will provide insights into the physiology of disease.

Such is a sample of long-term space activities being planned by NASA. These few items represent only a mere fraction of the long list prepared by space scientists.

In conclusion, space science during the latter part of the 1970s is not only maintaining, but gaining momentum as the most active of all man's sciences. The result is an explosion of new concepts, dreams, and plans. The brief discussions in this book have necessarily been limited to a mere fraction of the topics with which scientists are concerning themselves. Yet, all that has happened up to this point serves only as the foundation for the basis of space colonization and exploration during the '80s, '90s, and beyond. Close to earth, the planets will continue to be explored in hopes that we will gain a fundamental understanding of all the bodies in the solar system. The sun will remain the center of active studies, not only for its fundamental properties as the closest star to earth, but also because of its effect on the space environment. From the neutrinos and particles in the solar wind to the whole electromagnetic spectrum, scientists can study the sun in more ways than they can study any other object in the universe. The stars and galaxies are also under intense analysis. Of course, they are all much fainter than the sun, and so are much harder to dissect.

In any case, the expansion of man's senses to the humanization of space has just begun.

Glossary

Aberration of light: The apparent displacement of an object in the sky caused by the effect of the earth's motion on the direction in which light reaches us.
Acceleration: The rate of change in velocity of an object.
AIAA: American Institute of Aeronautics and Astronautics.
Anti-gravity: A hypothetical effect that would arise without the effect of the gravitational field of the earth or other celestial body.
Aphelion: The point at which a planet or other celestial object in orbit around the sun is farthest from the sun.
Apogee: In an orbit, the point of greatest distance from the center of the earth, the greatest altitude.
Asteroid: One of the 50,000 or more small celestial bodies revolving about the sun, most of whose orbits are between Mars and Jupiter.
Attitude: The orientation of a spacecraft determined by the relationship between its axes and some reference plane such as the horizon.

Black hole: Volume of space, only a few miles in diameter from which no matter or radiation, including light, can escape because of the intense gravity of an exceedingly dense collapsed star at its center.
Bolide: An unusually bright meteor or "fireball."

Cis-Martian: Pertaining to the space inside the orbit of Mars.
Closed ecological system: A system that provides for the maintenance of life in an isolated living chamber by means of a cycle in which the exhaled carbon dioxide and other waste matter are converted chemically or by photosynthesis into oxygen, water and food.
Comet: A luminous member of the solar system composed of a head or coma at the center of which a presumably solid nucleus is situated, and often with a spectacular gaseous tail extending a great distance from its head.
Cosmic dust: Small meteoroids of a size similar to dust.
Cosmic rays: Highly energized atomic particles that travel through the galaxy at speeds approaching that of light.

Declination: The angular distance, in degrees, of an object north or south of the celestial equator.
Deep space probes: Spacecraft designed for the exploration of space to the moon and beyond.
Diurnal circle: The circular path of a celestial object in its apparent daily motion across the sky.
Docking: The process of bringing two spacecraft into physical contact.

Eccentricity: The degree of flattening of an ellipse, or its departure from a circle.
Ecliptic: The apparent annual circular path of the sun among the stars.

Ellipse: An elongated circle.
Ephemeris: Computed positions of a celestial object for given dates.
Escape velocity: Minimum speed at which an object must move to escape the gravitational hold of another object.
EVA: Extra-vehicular activity.
Extraterrestial: Outside the earth.

Galaxy: System of billions to hundreds of billions of stars, along with nebulae and other interstellar matter, held together by attraction.
Globular cluster: Large, spherical group of stars within a galaxy.
Gravitation: Force of attraction that matter exerts on other matter.
"G": The rate of acceleration created by the earth, 32.2 feet per second per second.

Inclination: The amount by which the plane of an object's orbit is tilted to the plane of the earth's orbit.
INTELSAT: International Telecommunications Satellite.
Ionization: Process by which an atom loses or gains one or more electrons and becomes electrically charged.

Libration: An oscillation by which a planet shows portions of its usually hidden side; the libration of the moon permits us to see 59 percent of its surface.
Libration point: A point of equilibrium between the gravitational forces of two or more heavenly bodies.
Light year: Distance light travels in one year at a rate of 186,282 miles a second—about six trillion miles.
LRV: Lunar Roving Vehicle.
Luminosity: Brightness of a celestial object.

Magnetic field: Space around an object in which its magnetic force can be detected. The earth's magnetic field funnels charged particles from the sun into the polar regions to produce colorful aurorae, the northern and southern lights.
Magnitude: Relative brightness of a celestial body. The smaller the magnitude number, the brighter the body.
Meteor: Commonly thought of as a "shooting star." A small particle that enters the earth's atmosphere and is destroyed by friction before landing.
Meteorite: A solid body of iron or stone that falls on the earth's surface from outer space.
Meteoroid: A solid object moving through interplanetary space, of a size considerably smaller than a planet.
Milky Way: Home galaxy of our solar system; also, the band of light from stars and nebulae that arches across the sky.
Module: A self-contained unit of a spacecraft, either manned or unmanned, which serves as part of the building block of the overall structure.

NASA: National Aeronautics and Space Administration.
Nadir: Looking straight down from a satellite.
Nebula: Cloud of interstellar gas and dust.
Neutron star: Extremely dense collapsed core of a star that exploded; it consists almost entirely of neutrons.
NIMBUS: NASA Experimental Meteorological Satellite.
Nova: Star that violently erupts, briefly increasing its brightness by hundreds to millions of times.

Occultation: The hiding of one object in the sky by another, especially when the moon passes in front of a planet or star, or a satellite passes behind a planet.
Orbit: The path of a body or object under the influence of a gravitational force.

Parallax: The difference in direction of an object as seen from two places, or from two ends of a base line.
Perigee: The lowest point in an orbit around the earth.
Perihelion: The point nearest the sun in the orbit of a planet or object.
Perturbation: A slight change in the orbit of a celestial object caused by some outside force.
Planet: A major celestial body of the solar system.
Precession: The change in direction of the axis of rotation of the plane or orbit of a body when acted upon by an outside force.
Probe: A device inserted into space for the purpose of obtaining information about the cosmic environment.
Protostar: Concentration of interstellar gas and dust condensing to form a star.
Pulsar: Source of radio signals that pulsates in a rapid, precise sequence; generally accepted to be a spinning neutron star.

Quasar: Extragalactic starlike source of great radio power and extreme optical brightness.

Radiation shield: A wall made of lead or some other substance for the purpose of protecting bodies and instruments from the harmful effects of nuclear, cosmic and solar radiation.
Radio galaxy: Galaxy that emits its predominant energy in radio waves—hundreds to millions as much as the Milky Way.
Retrograde motion: Orbital motion of a satellite or comet which is opposite to the normal estward motion of almost all the members of the solar family.
Revolution: Motion of a celestial body in its orbit; motion about an axis usually external to the body.
Right ascension: Angular distance measured in hours or degrees from the vernal equinox eastward along the celestial equator.

Solar radiation: The total electromagnetic radiation emitted by the sun.

Solar wind: A stream of protons constantly moving outward from the sun.

Space: In common usage, that part of the universe lying outside the limits of the earth's atmosphere, "outer space."

Supernova: Massive star that explodes, releasing hundreds to millions of times more energy than a nova eruption.

Telemetry: The science of measuring a quantity of data, transmitted to a distant station from a spacecraft, which is then recorded, interpreted and measured.

Telescope resolution: Degree to which details of an image are made distinguishable.

Thrust: The pushing force developed by a space vehicle's rocket engine.

TIROS: Television IR Operational Satellite.

Trajectory: The path traced by any body, such as a rocket, moving as a result of externally applied forces. "Orbit" or "flight path" of such a vehicle.

Translunar: The space outside the moon's orbit about the earth.

Van Allen Belt: The zone of high-intensity radiation surrounding the earth beginning at an altitude of approximately 500 miles.

Weightlessness: A condition in which no acceleration or stress, whether of gravity or other force, can be detected by the observer within the system in question.

Zenith: The astronomical point directly above the observer.

BIBLIOGRAPHY

In the research and preparation of this book, I gratefully acknowledge having drawn upon a number of writings, particularly NASA and private aerospace company documents, research reports, and bulletins. As a selected bibliography, I offer the following titles which are of particular excellence. For convenience, they have been broken down into five main sections.

General

Ash, Brian, *Faces of the Future: The Lessons of Science Fiction*. New York: Taplinger, 1975.

Asimov, Isaac, *Is Anyone There?* New York: Doubleday, 1967.

Berendzen, Richard, ed., *Life Beyond Earth and the Mind of Man*. Washington, D.C.: NASA (Washington: U.S. Government Printing Office, SP-328), 1973.

Berry, Adrian, *The Next Ten Thousand Years: A Vision of Man's Future in the Universe*. New York: Saturday Review Press/ E.P. Dutton, 1974.

Boyle, Charles P., *Space Among Us: Some Effects of Space Research on Society*. Greenbelt, Maryland: Goddard Space Flight Center (NASA), 1973.

______ "Toward a 3-Dimensional Civilization," *Space World*, December 1970.

______ "In-Depth Exploration of the Solar System and Its Utilization for the Benefit of Earth," *New York Academy of Sciences*, **187**, p. 427, 1970.

Bracewell, Ronald N., *The Galactic Club: Intelligent Life in Outer Space*. San Francisco: W. H. Freeman, 1974.

Bradbury, Ray, "The Search for Extraterrestrial Life." *Life*, October 24, 1960.

Bundy, Robert, ed., *Images of the Future: The Twenty-First Century and Beyond*. Buffalo: Prometheus Books, 1976.

Cole, Dandridge, M., *Beyond Tomorrow; The Next 50 Years in Space*. Amherst Press, 1965.

Dalzell, Bonnie, "Exotic Bestiary for Vicarious Space Voyagers." *Smithsonian,* **5,** pp. 84–90, 1974.

Fimmel, R. O., et al, *Pioneer Odyssey: Encounter with A Giant*. Washington, D.C.: NASA/Scientific and Technical Information Office (Washington: U.S. Government Printing Office), 1974.

Hearth, D., *Outlook for Space,* Washington, D.C.: NASA (Washington: U.S. Government Printing Office, SP-387), January 1976.

Keosian, John, *The Origin of Life,* 2nd ed. New York: Van Nostrand Reinhold, 1968.

Lovell, Bernard, *Man's Relation to the Universe*. San Francisco: W.H. Freeman, 1975.

Lunan, Duncan, *Interstellar Contact*. Chicago: Regnery, 1975.

Maruyama, Magoroh, and Arthur Harkins, eds., *Cultures Beyond the Earth: The Role of Anthropology in Outer Space*. New York: Vintage, 1975.

Miller, Stanley L., and Leslie E. Orgel, *The Origins of Life on the Earth*. Englewood Cliffs, N.J.: Prentice-Hall, 1974.

Moore, Patrick, *The Next Fifty Years in Space*. New York: Taplinger, 1976.

NASA/Scientific and Technical Information Office, *A Forecast of Space Technology, 1980–2000*. Washington, D.C.: NASA (Washington: U.S. Government Printing Office), 1976.

Ponnamperuma, Cyril, *The Origins of Life*. New York: E.P. Dutton, 1972.

Sagan, Carl, ed., *Communication with Extraterrestrial Intelligence*. Boston: MIT Press, 1973.

______*The Cosmic Connection: An Extraterrestrial Perspective*. New York: Doubleday, 1973.

von Puttkamer, Jesco, ed., *Space for Mankind's Benefit*. Washington, D.C.: NASA (Washington: U.S. Government Printing Office, SP-313), 1972.

______ "On Man's Role in Space." Monograph, NASA Office of Space Flight, December 1974.

_____ "Developing Space Occupancy: Perspectives on NASA Future Space Programme Planning," *Journal of the British Interplanetary Society* (BIS), Vol. 29, no. 3, March 1976, pp. 147–173.

Space Industrialization

Bekey, I., H. L. Mayer, and M. G. Wolfe, *Advanced Space System Concepts and Their Orbital Support Needs (1980–2000)*. Aerospace Corp. Report ATR-76 (7365)-1, Vols. 1-4. Contract NASA 2727, April 1976.

Bekey, I., and H. L. Mayer, "1980–2000: Raising Our Sights for Advanced Space Systems," *Astronautics and Aeronautics* (AIAA). July/August 1976.

Ehricke, Krafft A., "Lunar Industries and Their Value for the Human Environment on Earth." *Acta Astronautica,* Vol. 1, pp. 585–682, 1974.

Glaser, Peter E., "Solar Power via Satellite," *Astronautics and Aeronautics,* August 1973.

Little, Arthur D., Inc., "Feasibility Study of a Satellite Solar Power Station." NASA CR-2357, 1974.

Martin Marietta Corporation, "The Viking Mission to Mars," *Today Bulletin,* No. 3, 1976.

McDonnell Douglas Astronautics Company-East, "Feasibility Study of Commercial Space Manufacturing." MDC E1400, 1975.

Minney, O. H., "The Utilization and Engineering of an Orbital Hospital," AAS Preprint 67-123, Dallas, Texas, May 1967.

Space Colonization and Lunar Bases

Asimov, Isaac, "After Apollo, a Colony on the Moon," *New York Times Magazine,* p. 30, May 28, 1967.

_____ "Colonizing the Heavens," *Saturday Review,* June 28, 1975.

_____ "The Next Frontier?" *National Geographic*. pp. 76–89, July 1976.

Chedd, Graham, "Colonization at Lagrangia," *New Scientist,* 1974.

Chernow, Ron, "Colonies in Space." *Smithsonian,* pp. 6–11, February 1976.

Clarke, Arthur C., *Islands in the Sky*. New York: NAL Signet Book, 1965.

_____ *Rendezvous with Rama*. New York: Harcourt, Brace, Jovanovich, 1975.

Dossey, J. R., and G. L. Trotti, "Counterpoint—A Lunar Colony," *Spaceflight*. p. 17. July 1975.

Heppenheimer, T. A., and M. Hopkins, "Initial Space Colonization: Concepts and R & D Aims," *Astronautics and Aeronautics,* pp. 58–72, March 1976.

Maruyama, Magoroh, "Design Principles for Extraterrestrial Communities," *Futures,* pp. 104–121, April 1976.

_____ and Arthur Harkins, eds., *Cultures Beyond Earth*. New York: Vintage Press, 1975.

Michaud, M. A. G., "Escaping the Limits to Growth," *Spaceflight,* September 1974.

NASA-Ames Research Center, "Space Settlements—A Design Study." Washington, D.C.: NASA (Washington: U.S.G.P.O., SP-413), 1977.

NASA/ASEE Systems Design Institute, "Design of a Lunar Colony." University of Houston, Rice University, NASA/MSC. NASA Grant NGT 44-005-114, September 1972.

North American Rockwell, Inc., "Orbiting Lunar Station (OLS) Phase: A Feasibility Study." NASA Contract NAS9-10924, MSC-02687, April 1971.

O'Neill, Gerard K., "The Colonization of Space," *Physics Today,* Vol. 27, pp. 32–40, September 1974.

_____ "Colonies in Orbit," *New York Times Magazine,* pp. 10–11, 25–29, January 18, 1976.

_____ "The High Frontier" and associated articles, *The Co-Evolution Quarterly,* Vol. 7, pp. 6–29, Fall 1975.

_____ Testimony in U.S. Congress, House Committee on Science and Technology on "Future Space Programs 1975." Washington, D.C.: U.S.G.P.O., 1975.

_____ "Space Colonies: The High Frontier," *The Futurist,* pp. 25–33, February 1976.

_____ "The Next Frontier—Space Communities," *Aerospace,* Vol. 13, pp. 2–7, December 1975.

_____ *The High Frontier: Human Colonies in Space*. New York: William Morrow and Co., 1977.

Parkinson, R. C., "Takeoff Point for a Lunar Colony," *Spaceflight,* September 1974.

Salkeld, Robert, "Space Colonization Now?" *Astronautics and Aeronautics,* pp. 30–34, September 1975.

Vajk, J. Peter, "The Impact of Space Colonization on World Dynamics." Lawrence Livermore Laboratory, 1975, UCRL-77584, and "Technological Forecasting and Special Change," 1976.

von Puttkamer, Jesco, "Developing Space Occupancy: Perspectives on NASA Future Space Program Planning." Proceedings, Princeton University Conference No. 127 on Space Manufacturing Facilities, Princeton University, May 1975 (in preparation).

Space Stations

AIAA Technical Activities Committee, "Earth-Orbiting Stations." *Astronautics and Aeronautics,* Vol. 13, pp. 22–29, September 1975.

Blagonravov, Anatolij A., "Space Platforms: Why They Should Be Built," *Space World,* Vol. H-8-92, August 1971.

Ehricke, Krafft A., "Beyond the First Space Stations." AIAA Meeting, NASA-Marshall Space Flight Center, Huntsville, Alabama, January 1971.

Hagler, Thomas A., "Building Large Structures in Space," *Astronautics and Aeronautics,* Vol. 14, pp. 56–61, May 1976.

Low, George M., "Skylab . . . Man's Laboratory in Space," *Astronautics and Aeronautics,* Vol. 9, no. 6, June 1971.

McDonnell Douglas Astronautics Co., "Manned Orbital Systems Concepts (MOSC) Study," NASA-31014, October 1975.

——— *Space Station: User's Handbook.* MCS G-0763, March 1971.

——— "Space Station Program Definition Study/Space Base." MSFC-DRL-140, Contract NAS8-25140, 1970.

——— "Space Station Systems Analysis Study." Part 1 Final Report, MDC G6508, Vols. 1-3. Contract NAS9-14958, (NASA/JSC), September 1976.

NASA-Marshall Space Flight Center, "Manned Orbital Facility: A User's Guide," Washington, D.C.: NASA Washington: U.S.G.P.O., 1975.

North American Rockwell, Inc., "Space Base." MSC-00721, Contract NAS9-9953, July 1970.

Parin, W., "Life on Orbital Stations," *Journal of Aerospace Medicine,* Vol. 41, no. 12, December 1970.

NASA/Scientific and Technical Information Office, "Future Aeronautics and Space Opportunities, *Space,* Vol. 1. Washington, D.C.: NASA (Washington: U.S.G.P.O.), 1974.

_____ "Outlook for Space: Report to the NASA Administrator by the Outlook for Space Study Group." Washington, D.C.: NASA (Washington: U.S.G.P.O.), 1976.

Proceedings:

AIAA/NASA Symposium on *Space Industrialization,* NASA-Marshall Space Flight Center, Huntsville, Alabama, May 26–27, 1976.

3rd Space Processing Symposium, *Skylab Results,* NASA-Marshall Space Flight Center, Huntsville, Alabama, April 30–May 1, 1974.

20th AAS Annual Meeting, *Skylab Results,* August 20–22, 1974.

U.S. Congress, House Committee on Science and Astronautics, *Space Shuttle–Skylab: Manned Space Flight in the 1970s.* Washington, D.C.: (Washington, U.S.G.P.O.), 1972.

U.S. Congress, House Committee on Science and Technology, *Space Shuttle 1975.* Washington, D.C.: (Washington: U.S.G.P.O.), 1975.

U.S. National Research Council, Space Applications Board, *Practical Applications of Space Systems: Materials Processing in Space.* Washington, D.C.: (Washington: U.S.G.P.O.), 1976.

von Puttkamer, Jesco, "Industrialization of Space—NASA Plans for the Next 25 Years," *The Engineer.* London, June 3, 1977.

_____ "The Next 25 Years: Industrialization of Space." 2nd International Congress on Technology Assessment," Journal of the British Interplanetary Society (BIS). July 1977.

Rockwell International Corp., "Space Station Systems Analysis." SD 75-SA-0301, February 1976.

_____ "Austere Modular Space Facility." SD 75-SA-0105, September 1975.

Smith, T. D., and Charhut, D. E., "Space Station Design and Operation." *AIAA Journal of Spacecraft and Rockets,* Vol. 8, no. 6, June 1971.

Index